SECRET MARS

The Alien Connection

M.J. Craig

Adventures Unlimited Press

Secret Mars

by Michael J. Craig

ISBN: 978-1-948803-30-4

Published by:
Adventures Unlimited Press
One Adventure Place
Kempton, Illinois 60946 USA
auphq@frontiernet.net

Printed in the United States of America

AdventuresUnlimitedPress.com

SECRET MARS

The Alien Connection

M.J. Craig

Other books of interest:

VIMANA
ARK OF GOD
BIGFOOT NATION
GIANTS ON RECORD
THINGS AND MORE THINGS
INVISIBLE RESIDENTS
ANCIENT TECHNOLOGY IN PERU & BOLIVIA
THE MYSTERY OF THE OLMECS
PIRATES AND THE LOST TEMPLAR FLEET
TECHNOLOGY OF THE GODS
THE ENIGMA OF CRANIAL DEFORMATION
ANCIENT ALIENS AND SECRET SOCIETIES
THE CRYSTAL SKULLS
ANCIENT ALIENS ON THE MOON
ANCIENT ALIENS ON MARS
ANCIENT ALIENS ON MARS II
ANCIENT ALIENS AND JFK
HIDDEN AGENDA

Table of Contents

Preface

This book is about a search for the truth. My reason for writing it is because I want to know if the human race is alone in the universe in which I live. I need to know because I want to understand who and what I am, and my place in the scheme of life. I want to know why I am walking around, with these conscious thoughts and imaginings in my head; seeing, breathing, feeling alive. Who am I? Why am I here? What is my purpose?

From time to time, every human being on Earth will share these thoughts and this deep yearning to know, but there are few answers set in stone. Science, philosophy, and religion will claim to offer the knowledge we seek, but not one of those human oracles will be able to present to us the whole truth and nothing but the truth. And yet in order to discover who and what we are, we have to depend on them to teach us what they know and to show us the way, the best they can.

We can ourselves become aware of much through our studies. We can piece together the evidence and come to conclusions. But upon crucial details that affect the whole picture of truth, eventually, we must rely on someone else to provide the facts from which we will draw our opinions and form our beliefs. We need and depend upon learned scholars and scientists who specialise in the specific subjects we need to know about.

These are the gatekeepers.

In their highest form, they hold up a beacon for mankind, proclaiming, "You may safely walk down this road, because I have lighted the way for you" and we must trust to their judgment, their expertise, and their honesty, in wishing to serve mankind to the best of their ability and intelligence.

Thankfully, of course, there are many such trustworthy individuals who, fuelled by their passion and desire to learn the truth and with wonderful humility, eagerly pass on the treasures they have discovered for the benefit of the human race. And for those souls, I have the utmost respect and trust.

But then some are less so inclined. These are gatekeepers who, for whatever reason, have decided that it is not in their interest to have the majority of the human race know the truth about a matter of great importance. Sadly, it is something we all know and recognize about our imperfect world. For, with monotonous regularity, stories ever continue to emerge in the world of politics and business, exposing some person or other in authority, manipulating events for their greed, or from a position of fear.

Money and the accumulation of wealth and power are at the heart of corruption on our planet and of that fact, no aware citizen can be in any doubt. It is to be hoped of course that over time, humanity will be able to evolve to a state where people no longer need to behave in such a way, but for now, on our planet today, this is what we have; people in positions of world authority, crafting agendas for personal gain, profit, power, and control.

What surprised and then angered me and eventually led me to write this book, was when I began to suspect that this kind of behaviour extended to matters that I would have hoped were far beyond the reach and scope of lesser evolved people; matters that were global and of deep spiritual significance and consequence. Profound questions that involved who we are and where we come from as a race of conscious beings, living in a vast mysterious universe: our vital research and discoveries into the wonders of space and on other planets and of course the question:

"Is the human race alone in the universe?"

Could anyone have any reason for wanting to suppress and keep quiet discoveries in this revered domain? But evidence has sadly emerged that this is indeed what has been happening and that it's been going on for a very long time. I realized that there were restrictions and barriers in place; odd, gaping holes that should be linking one arm of research to another so that commonsense, logical investigation can be done and our collective, planetary knowledge expanded; people in responsible positions closing doors and refusing to open others.

And there was one subject above all that drew the most peculiar response and provoked the most alarmingly illogical and often infantile reaction:

The matter of extraterrestrial intelligence: the possibility of beings from another world, having had contact with humanity and our solar system.

And it so happened that the planet Mars became the focus for the problem I was facing. It seemed to me to present a stage that portrayed all the things that were not being done right in mankind's search for knowledge, in particular, the search for and question of there being life elsewhere in the universe.

The first alarm bell rang when I noticed a strange reluctance from the American space administration NASA to engage with the subject of "The Face on Mars", as it came to be known back in the 1980s and 90s. There was a potential artifact from an alien civilization, and yet, instead of an excited, open-minded scientific investigation of the anomaly, all we saw from this public-funded agency was a complete dismissal and indeed a very unscientific rejection of even the *possibility* that it might be real evidence of ET.

Then later, when more advanced spacecraft began approaching Mars with far superior cameras, I observed a stubborn refusal by NASA to even guarantee that new pictures of The Face would be acquired so that science could verify if it was indeed just a funny shaped hill or something of far deeper significance. It seemed to me that 'science' just did not want to know.

Then in 1999, I saw a startling new picture of flowing liquid water on the planet Mars, discovered by an independent researcher who was studying images from NASA's database. I wondered why I didn't see this news splattered across the pages of the world's media. Not a whisper. Even stranger than that, some thirteen years on, NASA was still making decisions to land their expensive spacecraft in areas far away from where this water was found. Incredibly, they seemed only interested in studying sites on Mars where water used to exist, millions of years ago!

I mean, correct me here if I've missed something. If anyone with a degree of commonsense was in charge of a mission to search for life on Mars, surely they would have directed their spacecraft to areas where water was known to exist, not to places where there wasn't any? So why didn't they do it?

I then found that NASA hadn't sent an astrobiology experiment to Mars to detect life since the Viking landings back in 1976. Now, why was that? Five spacecraft have landed on the Red Planet since Viking and they have all been geology missions, sent to study rocks and soil. Not one of them with the ability to detect the presence of life, including the latest one, Curiosity.

Here on Earth, we have 20,000 children dying every day from starvation and we spend a billion dollars to go look at some rocks on another planet? If we are going to spend that kind of money we had better damn well justify it by seriously expanding the knowledge of the human race, not by sending a spacecraft all the way to Mars just to touch a rock with a robotic arm—and satisfy the indulgence of a handful of planetary geologists.

I had presumed like most people did, that the purpose of spending billions to go to Mars was to search for life. It confounded my sense of reason therefore, that decisions had been made to avoid landing a craft where water had been seen. Every biologist will tell you, "In every drop of water you will find life". We are much more likely to find evidence of microbes in Martian, water-saturated soil than in a billion-year-old rock found on planet Earth, on an icy, barren plain in Antarctica. And remember the media frenzy *that* caused…

And here we are today and the sensible thing is still not happening. Spacecraft after spacecraft is sent to Mars at a huge cost, but the most important question about Mars has still not been answered. Go to the NASA website and ask, "Is there life on Mars?" The reply remains the same as it did over thirty years ago, which is:

"We don't know if there is life on Mars yet".

That is quite a sad return of knowledge on over fourteen years of continuous scientific investigation of our planetary neighbor and at such enormous investment. It just does not make any sense…and that's why this book has been written, to try and make sense of it and ask some questions.

If you smell a rat you have to follow the trail. After all, their priority may not *be* to find life on Mars. Or, they may already have found it and don't want to tell us about it yet. If so, then why not? Why delay? Are they procrastinating by feeding us other science missions that they manage to convince us should be done beforehand? Or, that in their opinion is just as important as the search for life? They do seem to be stretching it out a bit. I mean, planning a rock sample return in the year 2020?

Perhaps they think they can get away with it by sending us all wonderfully dramatic pictures of Martian geology or focussing instead on the possibility of finding signs of primitive *ancient* life, microbes, and the like. Or, present-day 'habitability' as the latest mission Curiosity will doubtless confirm. Anything it seems but proving *present-day* life exists on Mars, right now.

Are we entitled to make such a radical claim? An accusation which now falls into 'conspiracy' territory? Yes, I think so. In light of the lack of real progress in answering the most important question of all concerning Mars, I think we are more than justified, and I believe the evidence in this book will support this line of enquiry.

But how can we question the dedication and expertise of some 18,000 employees and scientists at NASA? They can't be hiding something, as the organization is far too big and too public and everyone would, of course, agree with that statement. No, for such deliberate and long-term manipulation of a national space program you would have to go to the policy makers, and above, where decisions are made far away from the public ear.

I am not a scholar, I'm a layman. I have a deep interest in this subject but I have no specialized knowledge. I have therefore needed to draw upon research done by others to support the position I have taken here. Oh yes, some have been pursuing this matter for many more years than I have and who have been challenging the deaf ears of mainstream science to wake up and be brave, to little avail, since the 1980s.

There has been serious work done by Hoagland, DiPietro, Molenaar, McDaniel, Carlotto, Van Flandern, Crater, Torun, Brandenburg, Erjavec, Pozos, Skipper, Filotto, Hain, Isenberg, Palermo and many others in this field, whose persistent and stubborn research has made this book possible and to whom I offer my gratitude.

Fundamentally I believe that something serious has gone wrong with the scientific rationale in this world in the search for extraterrestrial life—especially *intelligent* extraterrestrial life —and this unfolding story on Mars is but a symptom and result of the warped and blinkered vision that blights this subject.

Perhaps it is because a new paradigm is emerging, throwing up conflicts of interest in the halls of power in science, religion, politics, and business. All I do know is that the potential discovery of life and intelligent beings in the universe apart from ourselves will be a monumental one, with huge ramifications for the human race and our society, and as such demands a rigorous and challenging appraisal of those who are studying this subject on humanity's behalf _and of those who are not but who should be.

I hope my readers will forgive my oft brash (I know) analyzes and demeanour throughout this book's narrative, but all I'm trying to achieve here is to try my best to reveal the truth to you, whatever it is, with all the passion, insight, reason and intuition I possess.

It is sometimes difficult for me to curb my annoyance with the sense of injustice and dismay I feel, that some arrogant, misguided persons have seen fit to feed me and the rest of humanity a lame, inaccurate and false view of life in the universe while they selfishly ponder in secret what to do with the real and astounding jewels of truth that I am certain have been discovered during our missions into space and from our distant observations from the Earth.

I do not take kindly to people deliberately depriving me of information that could throw light upon my existence as a human being. This, to my way of thinking, is a heinous crime. If you feel the same way I hope you consider my effort here to be a worthwhile attempt to shake things up a bit.

M. J. Craig

The West Country, England

Author's Notes on the Second Edition

Since the first publication of *Secret Mars* in 2013, new scientific discoveries concerning the Red Planet continue to be made, the most notable of which came in September 2015, when NASA finally decided to tell the world that liquid water flowed on Mars, even though they had made this discovery some sixteen years earlier in 1999.

What has not been announced however is the increasing evidence indicating an intelligent civilization used to live on Mars. But the thought occurs, "Well, if it takes NASA sixteen years to let us know about liquid water, how long will it take before we get to hear about alien artifacts?"

In the meantime, we can't wait for the sloths of academia to stir, or the analysts whose job it is to study the new images from Curiosity, to remove their geologically tinted spectacles that see only rocks. We must push on and gather more evidence for those who are ready to see it.

The purpose of this new edition, therefore, is to simply expand upon and improve the evidence casefile. To this end, several new pictures from the Curiosity Rover have been included that strengthen the case for a technological civilization having once inhabited Mars. I have also added images that suggest Martian fossils have both been discovered, and ignored by NASA science teams.

Some new material has been added too, the most prominent of which discusses the 'Planetary Protection' problem that NASA is experiencing in the search for life on Mars, while I've also edited some passages, made some corrections and improved the overall presentation of the images.

All in all, this is a superior edition of my book that I hope will be welcomed and appreciated by those intrepid souls who are not afraid to engage with this subject, and who, like this author, is simply interested in discovering the truth, whatever it is.

Acknowledgements

Image credits

Planetary images presented in this work have been sourced from official science data made available to the public through NASA/JPL (The National Aeronautics and Space Administration/Jet Propulsion Laboratory) and associates, and the ESA (European Space Agency). Images from the MOC (Mars Orbiting Camera) on Mars Global Surveyor were provided courtesy of Malin Space Science Systems at www.msss.com.

Image Sources

My thanks to the following for permission to use their images in this book: Dr. Mark J. Carlotto; Samantha Joye; Tim Beech; The Oriental Institute of the University of Chicago; The Sedgwick Museum of Earth Sciences, University of Cambridge; Cornell University; Arizona State University; Caltech; Max Planck Institute; Landsat; Google Earth; Digital Globe; ESA; DLR; Prof. G. Neukum, FU Berlin; NASA/JPL; Malin Space Science Systems.

It is imperative that knowledge funded and acquired by the state is shared openly with the public and the science community to enable transparency and trust. I am therefore grateful to the American and European space agencies and their affiliates for making available to the world the images acquired by their spacecraft.

Each image is credited to the agencies responsible for producing it, as far as I hope my research has been accurate. Corrections are most welcome. I have also included a link to the web address for each image I've used as evidence material so that the reader can view and study the images for themselves, which is an important part of this project (see image index). This is especially crucial too because much more detail will be available by viewing the original online source images.

Preparation of images has been done solely by the author except where stated. These are mainly cropped versions of larger files that exist on the NASA and ESA websites, so please bear in mind that when you follow the source links to view an original image, you may have to do some scrolling and scrutiny before you find the actual object featured.

Image Discoverers

Although NASA and the ESA have provided virtually all of the images in this book, it has certainly not been they who have drawn attention to the mysterious objects that are examined throughout these pages. The important finds in this book can virtually all be credited to independent scientists, researchers, and individuals. Some of these have written books, papers, articles, or host websites or social media pages specialising in Mars research, and it's to these dedicated and persistent souls we owe our respect and thanks.

In particular to the pioneering research of Richard C. Hoagland, whose remarkable work over the last 25 years at *Enterprisemission.com* has revealed several high calibre discoveries that warranted prominent coverage in this book. As too the extensive scrutiny of Mars images by the late J. P. Skipper at *Marsanomalyresearch.com*, whose research provided several fascinating images that still provoke debate. Of no less significance has been the extraordinary work done by Rami Bar Ilan at *Exclusive Mars Images*, whose relentless combing of the Curiosity pictures over recent years has unearthed a treasure trove of possible Martian artifacts that may never have seen the light of day were it not for him.

I offer my sincere gratitude to all who have contributed their time, dedication, tenacity, and sacrifice to identify and share their extraordinary discoveries with the world in an effort to uncover the truth for all of humanity. I hope you consider this book to be a worthy testimony to your work.

Can NASA Images Be Used as Evidence?

So, what about the actual images presented in this book? As I will be suggesting that some of them may show evidence of a previous civilization that has lived on the planet Mars, you may ask, "Can these pictures really be taken seriously?" Ultimately that will be for you the reader to decide. However, I should in fairness warn you that the most common arguments heard against the pictures shown in this book, are:

Geology—there is absolutely nothing artificial shown in these NASA images. All they show are unusual rocks shaped by geological, erosive forces, such as ventifacts caused by wind-blown sand.

Fantasy—any resemblance to broken machinery or a common human object such as a plate, box or hinge, is just the result of our tendency to see familiar things in nature such as 'faces in clouds'—pattern recognition, simulacra, and tricks of light and shadow that give rise to pareidolia and optical illusions.

All of which are very valid and critical observational criteria upon which to analyze a strange object in a picture. And I would stress too, that anyone who claims to see a 'tool wrench or teacup' in an image from Mars without first having considered these rational alternatives, is probably going to delude themselves to a very large degree.

So I can, therefore, assure you that faced with such a barrage of all too rational alternative explanations for the strange objects I'm going to present to you for consideration, I have consequently taken some care to include only those pictures that I think are worthy of being labeled, "Possible evidence of an extraterrestrial artifact".

Yes, most scientists and academics will certainly dismiss every image in this book as pure fantasy and will not take the hypothesis put forward at all seriously. They will generally argue—as do the NASA rover mission teams—that as the rovers trundle along the Martian surface taking pictures every few seconds, they literally pass by millions of rocks and that given those huge numbers, we are certainly going to come across a few that are going to look very strange, some probably resembling a cube or a pyramid or some other very geometric form—perhaps even a coca-cola can. And who indeed can argue with this rationality. Not I certainly and neither will most reasonable people.

However, the only problem with this argument is that it is based upon a possible flawed assumption, which is: that the planet Mars has never been inhabited by an intelligent species in its entire history. They reason that because Mars has never been inhabited, then anything that looks like an artificial object can only *ever* be a rock—which of course, if, no intelligent species *had* ever stepped foot on that planet, then their logic would be entirely inescapable.

It is believed by science that no intelligent life could ever have evolved on Mars during her entire 4.5 billion year history, but even if that should be correct, it does not mean that intelligent visitors from another world could not have stepped foot onto the Martian surface. And the moment we entertain the idea that Mars may have been visited at some point in her long history by an intelligent species—a civilization that may quite reasonably have set up bases, settlements, industrial facilities, etc.—then the analytical filters will of necessity have to be reset: that 'shiny, cube-shaped stone' may well be a *metal box*; that 'strange rock with regularly spaced spokes', may indeed be a *gear wheel*; that 'optical illusion which seemed to show a small set of wheels on an axle' could be . . . *a set of small wheels on an axle.*

As such, I believe it to be entirely reasonable to keep an eye open for possible alien artifacts on the surface of the planet Mars, and indeed on all the other planets and moons in our neighborhood too. In fact, given the growing realization in science that millions of planets in our galaxy could harbor life—which is one heck of a lot of planets—I think it would be quite irrational not to do so.

Also, this awareness of the likely extent of life in the universe should be expanding everyone's mind to the probability that there will be found to be other civilizations in space who are way more advanced than humanity, maybe thousands of years older than us, and most importantly—who would have discovered interstellar space travel a very long time ago. And it is in this light that I believe we should be taking seriously the idea of finding alien artifacts on Mars, for we have only to imagine a race of space-faring intelligent beings who, with a mind for exploration and colonization just as we have today, simply deciding to visit our solar system.

A million years ago, their telescopes or long-range scans might have revealed two 'goldilocks zone' planets bearing oceans of water (Mars and Earth)—exactly as we are today assessing Earth-sized, possible water-bearing exoplanets that orbit distant stars. They would have landed, maybe stayed a while, maybe for a very long time. Either or, they surely would have left some trace of their presence on the surface and we should not at all be surprised therefore to discover such remnants on our subsequent travels to that planet. Indeed an official report commissioned by NASA back in the 1950s, expected as much *(see chapter 1, "Brookings Report")*.

Interestingly though, some of the most vociferous challenges I've experienced come from people who are genuine Mars enthusiasts and lovers of science. They appear to revere Mars as a virgin untouched planet in our solar system that no conscious being has ever set foot upon—a pure, pristine world ripe for exploration and scientific study. They don't want to hear talk about 'aliens' or 'ancient civilizations' to spoil and blemish this romantic vision they have of Mars, proclaiming that Mars is wondrous enough without the need for outlandish ideas and conspiracy theories.

And yes, Mars is absolutely a wonderful and great mystery to behold. I'm one of those peculiar folk too, in awe of Mars and who wants nothing more than to walk upon the surface of this world, and I know I'm not alone.

But above all, it's the truth we want to wake up to, not a dream. As things stand NASA does not consider the archaeological search for alien artifacts on Mars, to be a serious scientific subject to consider, and consequently expend zero resources on exploring this possibility. I am left to wonder how history will judge them on that stance.

Introduction

The internet is changing our world to a profound degree. The free flow of information that can instantly be shared by millions of people across the planet is revolutionising freedom, democracy, and the accountability of governments, institutions, and corporations to the people of this planet.

The information presented in this book is one such blessing of this revolution, as the American and European space agencies have made available to the public many thousands of images taken by spacecraft of their explorations of Mars. As these expensive missions were funded by you and me through our taxes, this is exactly the way it should be and is all well and good.

Furthermore, these agencies have also actively encouraged the public to become involved in studying these images, as well as the scientists—and they have done so with great enthusiasm and vigour. However, most people want to know if there is more to Mars than rocks, ancient signs of water and evidence of microbial life, as NASA-funded scientists are pedantically concerned with and so have been examining the images to see if they can make far more significant discoveries than the scientists have been able to.

And as this book will show, they have done.

Individuals, in their spare time, after a day's work at the office or in the factory or warehouse, have found some incredible evidence that NASA scientists won't even officially discuss, investigate or talk about on the news. The people have found evidence for nothing less than *an ancient civilization on Mars.*

The images you will see throughout these pages are an extreme provocation to those scientists who believe they are the only ones qualified to know what should and should not be pursued as valid science, in the investigation of our planetary neighbor. Yet in 1999 the public discovered that liquid water flowed on its surface and asked:

"Why does NASA not land its spacecraft where this water is?"

. . . *But NASA scientists continued to land their craft in bone dry, boring sites where there is little chance of finding life.*

The public went on to discover pictures that showed possible signs of artificial objects lying in the Martian sand and asked:

"Why are there no archaeologists on the Mars research team to study these images?"

. . . *And NASA dismissed and ridiculed the idea that any intelligent life could ever have existed on Mars.*

So, while NASA ponderously rolls out their legitimised scientific program to analyze Martian geology and to search for evidence of microbes and ancient sites where water used to exist, the public is looking for *larger* signs of life in the pictures coming back from the Red Planet. Why? Because evidence of intelligent life on Mars, past or present, is a million times more interesting and relevant to the human race, than the geological history of the planet—which appears to consume most of the scientists, the spacecraft, and the dollars of NASA's Mars exploration budget.

What the public has found is revealed here in this book: independent researchers who have been doing the work that the space agencies have been ignoring. It is the ordinary people—not NASA scientists—who have revealed possible evidence that Mars was likely to have been inhabited by an intelligent civilization at some time in its history.

NASA will probably one day catch up with all of this research, or they may already know about it but are staying quiet for now, but whichever way this plays out in the future, this volume honors today the work being done in the world by researchers who, inspired by their search for the truth and empowered by the internet, have analyzed images taken by the American and European space agencies and made possible *world-changing discoveries*—discoveries which those agencies to this date have not acknowledged or declared to the world.

We're talking about visual evidence—pictures taken by our very own imaging spacecraft—that show the possible remains of an ancient civilization on the planet Mars.

This is a momentous claim to make and as such demands vigorous follow-up action and research. We need all the images of Mars to be extensively scrutinised. Preferably, we need experts who are highly qualified to analyze and examine them—but if the academic and scientific community won't look at the pictures and challenge the mainstream, then they leave the people with no choice but to do it themselves, as best they can.

The fact is that NASA and the ESA will openly discuss the geological discoveries made on Mars and the possibility that Mars may once have had or may still have primitive microbial life, but they have expended absolutely no scientific resources whatsoever on examining the potentially far more important discovery for the human race—*the possible existence of archaeological remains.* It is just not on the table for discussion and after considering the evidence presented here in this book I trust you may consider that it should be.

In a world where most of its inhabitants are living on the edge of poverty and starvation, it beggars belief that we are wasting valuable Earth resources and billions of dollars to send robotic spacecraft to Mars, only to learn something about that planet's rocks, soil, and microbiology. If we are going to spend vast sums of money going into space at all—and I believe we should do—it can *only* be justified by presenting the human race with a rich discovery of knowledge concerning our place in the universe. Not to simply acquire data on powdered rock samples so that we can understand the details of how a planet formed billions of years ago. It is an immoral and abysmal waste of precious resources.

The search for life—intelligent life—should be the most important reason for going into space. This book will show you that currently, it is not. In fact, the only direct search NASA and mainstream science are making for intelligent life in the universe is a short-sighted project run by the *SETI Institute.*

SETI uses giant telescopes to look for radio signals from far distant stars. They sit there night after night, year after year, desperately hoping there might be a civilization in outer space that's using radio to communicate between the stars. This, despite the fact, that with radio waves traveling at the speed of light, it will take many years to exchange messages—not a very practical way to say "Hello, how are you?" and then have to wait years for an answer. 'SETI' stands for "The Search for Extraterrestrial Intelligence", which for such a scientific and intellectual lame duck, one might feel to be a galling misnomer.

This book will show you that we don't need to go looking for signs of E.T. many light-years away, because there is plenty of evidence for its existence, or it having once existed, in our very own solar system right now—and that some of that evidence still exists today on the planet Mars. And this reality must raise a serious question for the inquisitive and rational mind:

"If signs of intelligent, alien life exist on Mars, why are the space exploration authorities in both the United States of America and in Europe not investigating this evidence? And how much are our leaders aware of this strange omission in scientific research that their expensive national budgets are paying for?"

To find proof of intelligent, extraterrestrial life will be the greatest discovery in the history of the human race, but could there be a *problem* with such a monumental discovery? Are there people who are *afraid* of having that question answered? Or who, knowing the truth already, do not want you and me to know about it? Is there some deep issue at the political or some other level, perhaps religious, that is restricting the free release of a great and incredible secret?

It may well be that the burrow goes very deep, but something has to give. We have seen the internet empower people the world over to bring oppressive governments and rampant corporations to heel. If there is a secret to be revealed on this subject perhaps now is a good time to discover if our political leaders and institutions in the so-called civilized West are truly subservient to the people, truth and democracy, or who perhaps serve and are more motivated by other more clandestine factors.

I hope that the future will deplore and dissolve "Truths protective layers", as the late Apollo 11 astronaut Neil Armstrong deftly put it, and the human race will be able to go forth into the wonders of space, inspired purely by the blazing torch of humanity's noble search for truth, knowledge, and wisdom—not left staggering half-blind into the abyss, twisted and constrained by corporate contracts, secret society agendas, and other puerile, unevolved motives.

When scientists eventually acknowledge for themselves the evidence that exists in this book and thereafter great strides are made in our understanding of the planet Mars regarding both its past and present, I hope they will have the good grace to credit the open-mindedness, vision, courage, and tenacity of those not of the mainstream science communities, who sought for many years to break down the doors that have held back truth, knowledge, and understanding of this world—and of the real universe in which we live.

I'm speaking of the independent Mars researchers from world society; the internet 'rogues, charlatans and advocates of the devil', who challenged the veil of secrecy and deception that currently envelops our 'advanced civilization' here on planet Earth. It has been their diligent, tenacious, and dogged work that has brought to my attention the many controversial images of possible Martian water, life, and signs of civilization shown in this book. They, who having witnessed and bore the frustration of NASA's lame obsession with rocks and geology, just went off to search for themselves the real jewels of knowledge that Mars has to offer.

This book is about *People Power*. It is about the old order crumbling away as a new conscious, evolving world vision continues to dissolve a control over human society that has for centuries lain in the hands of elitist families, esoteric orders, and self-obsessed tyrants. To deliver the blueprint for humanity back to where it belongs: in the hands of the communal, planetary awareness for all to share and prosper.

In preparation for this book I listened to several presentations featuring planetary scientists working for NASA and the Mars Exploration Program and what struck me was the passion, enthusiasm, and eloquence displayed by every one of them for their particular scientific niche, which I found heartening and inspiring.

As I listened I felt myself deeply respecting their dedication and discipline to having attained such highly crafted expertise, but afterward began to feel a dispirited sense of sadness too, that these wonderful and talented people were being cheated of much greater discoveries that they should now be making on behalf of humanity.

The evidence in this book will I hope, have them up in arms and pounding on their director's door, but maybe some will just feel the need to return to their laboratories and continue as they were, feigning ignorance of what is revealed here. We shall see.

No doctorate or scientific accreditation is supporting this book. I am a layman looking over the shoulders of more highly qualified intellects, trying to grasp the larger picture of something that I feel has gone seriously wrong with our exploration of space. Maybe someone like me and the thousands of others outside of the inhibiting, daily minutiae of engrossing detail that must absorb NASA employees and their contractors, can perhaps see something they cannot, or just will not see.

The only criticism I have of science is when alternative explanations are refused consideration; when judgments are formed which have a bias against a particular hypothesis even though rational and credible evidence offers support to it. Scientists are of course only human, like all of us, but as gatekeepers in a highly privileged position, they also have a responsibility to keep a seriously open mind, surely knowing what the price will be for keeping it closed.

Yes in this fast-moving world where daily discoveries reveal an ever more fascinating universe before us that stretches our imagination and ability to understand it, and where the depths of space now reveal and confirm to us that many Earth-like planets probably orbit other stars, and from where, even on our little blue world, we sketch out plans for colonizing a planet in our very own solar system—even before we can actually send someone there—we nonetheless will not take seriously, or even put on the table for discussion, the one very likely, logical scenario that will explain much of the evidence presented in this book.

This is real evidence that NASA claims does not exist. Which is strange, given that most of it comes from NASA's own library of planetary images acquired by their orbiting spacecraft and landers. Do they not look at them? Or only the images that fit into their current well of thinking?

It must be the time of Galileo again, where the old must fall before the new. Of the time when the gallant Bruno fell in flames before the ghastly closed-mindedness of an authority that sought to preserve the way, things were at all costs, even at the expense of truth.

Let's see who will recognize themselves, sitting safely snug amongst the crowd or standing alone far away from it. There is a new paradigm emerging, and only a tiny few scientists have so far stepped beyond their comfort zone to embrace its coming. But then, they should know more than most this has always been the way of it.

To the scientific community, let's see where you choose to stand. To everyone who is not a scientist, I hope you will have an easier time assimilating the information revealed here.

I believe this book will show you clear signs that intelligent life once inhabited Planet Mars. The evidence will also point towards the possibility that Mars did not evolve an indigenous civilization, but rather instead that it was once colonized at some point in time by intelligent beings from beyond this solar system. And thus the possibility must remain, that their ancestors may still be there.

Will this evidence be enough to alter the priorities of the American Mars exploration program? Will they send a mission to search for ancient ruins? Are they prepared to make discoveries that will change the world as we know it and change forever our view of life on Earth and elsewhere in the universe?

We confidently prepare for the day when such questions will not bring about fear but instead open minds and hearts, and enlightenment for all humanity.

Chapter 1: Obfuscation

My reason for writing this book is that I believe the Americans have discovered signs of an intelligent civilization that once existed on the planet Mars and have decided not to tell the world about it, at least not yet. This discovery is being kept a secret. I want this to change. I want the truth to be revealed so that every person on our planet will learn all about it and can begin to absorb the incredible implications this discovery has for both themselves and the world at large.

I believe that such profound knowledge belongs in the hands of the entire human race so that we as an intelligent species can collectively begin to contemplate the awesome significance of this revelation and so reassess our place in the universe. It does not belong in the hands of a select few to decide which truth will be fed to humanity and which will be denied.

I do not say that 'NASA scientists' have discovered this evidence, for if they had done so it would surely be impossible to hide such a discovery—although fear of announcing it would certainly exist in some circles. No, I'm talking about some shady program, certainly not public, that has achieved a covert study of the planet Mars, found evidence of intelligent life, and has decided to keep the truth from the world until they see fit to reveal it, if at all.

This secret agenda would probably require the cooperation of senior NASA officials, but who knows. However, as NASA has generally towed a non-inquisitive and suspiciously unimaginative and unscientific line over certain aspects regarding the exploration of Mars and represents the world's public face of space exploration, NASA itself will be deemed here as mainly to blame.

Essentially and disturbingly, it appears that this substantial influence has managed to direct a space exploration policy that has successfully *prevented* NASA scientists from searching for life more vigorously on Mars and discovering much more than has so far been revealed by the current NASA space missions to the Red Planet.

We are not talking about the discovery of microscopic, primitive life which will sooner or later be announced to the world anyway, but the far more significant omission from NASA's so-called search for life on Mars: the complete avoidance and non-discussion of the signs of *intelligent* life, which our evidence will show has almost certainly existed on Mars in the ancient past and which may even still be there today.

The covert agenda which I believe exists is more than happy for mainstream scientists to be absorbed with their pedantic studies of Martian atmosphere, past climates, geology, ancient sites of water and the search for microbial life on Mars, for this book believes there to be a *two-tier* investigation of Mars:

- **One study for the public** . . . the meager results of which we get to see on our TVs and the web circulated by the news media and by NASA.

- **Another study being done in *secret*** . . . the one that you and I don't get to hear about, that is probably classified way above top secret. Whether even the American president gets to see this study may be worth a thought.

What I'm saying will be supported by the evidence presented in this book. However, whether you consider the evidence to be sufficient grounds to challenge the world view you have come to know and accept will be entirely up to your good judgment. I know this may all seem like an extraordinary claim to make, but I hope to show that it is nonetheless entirely plausible and is true.

At first, this accusation of NASA and American secrecy may not make any sense and you would be forgiven for dismissing it out of hand. I mean, why would the very organization and scientific body that is exploring space on behalf of the human race and above all, actively searching, we presume, for life on other worlds than our own, want to *hide* such a fantastic and world-shattering discovery?

Well, perhaps the notion of 'world-shattering' might be the problem here; the potential crux of reasoning that might dissuade authorities from releasing such knowledge and discovery. For, if you release information that is going to change the world, then someone somewhere is surely going to ask what will the consequences be, and just *how* will the world change exactly?

For centuries our planet has been run in a certain way and there will be those who may not want that system to change very much because they may lose what they have gained. This research suggests there is much that is being kept hidden from the public and that the consequences of revealing the truth about extraterrestrial matters are indeed far-reaching and not to be underestimated. I believe that some people who hold powerful positions have made that calculation and decided that this information must be controlled.

I accept that this is something that by its very nature is going to be very difficult to prove, so we won't attempt to do that here directly. What we will do, however, is to lay out the evidence and trust that with an open-minded observation of known facts, visual evidence in the form of official NASA and ESA images, and above all the power of reason, that on balance the reader will become aware there is something seriously wrong with the official picture we have been given concerning the reality of life in our universe, our solar system, and regarding the current human exploration of our planetary neighbor, Mars.

The possibility that the Americans would want to hold back such an enormous and world-changing discovery is a deeply disturbing one, and the potential consequences and ramifications for such deception are huge. And maybe it would be unfair to pile the whole blame on that nation's shoulders, in as much as a deception of this magnitude may well be found to be sourced to an international accord and agreement to which America is but one signatory. It may just be so that because the United States of America has the most advanced space program in the world it now finds itself in the firing line.

We will see.

Avoiding the Search for Life on Mars

So, if the reason for holding back the discovery of life on Mars is because there are some people who fear the big changes the release of this knowledge will bring to the world, is there any evidence that the search for life on Mars is being delayed, curtailed or restricted in any way?

1. (L) Probable Martian meteorite: ALH84001
2. (R) Possible fossilized Martian nanobacteria from ALH84001
Image credits: NASA/JPL

I long believed that if they were to find life on other planets, the Americans would fall over themselves with joy to want to excitedly announce the discovery to the waiting world. Well, they kind of did that with the discovery of possible microbial evidence for life on Mars in that little 9-inch meteorite they found in Antarctica and which prompted the then U.S. President Clinton to make a televised announcement to the world in August of 1996.

NASA scientists had claimed it had been blasted off from the surface of Mars around 15 million years ago following a meteorite impact on that world and that some 13,000 years later, after a long journey orbiting the solar system, it had landed in Antarctica where it was found in 1984. An electron microscope later revealed what are possibly fossilized remains of Martian bacteria, but this evidence has since been challenged although not discredited.

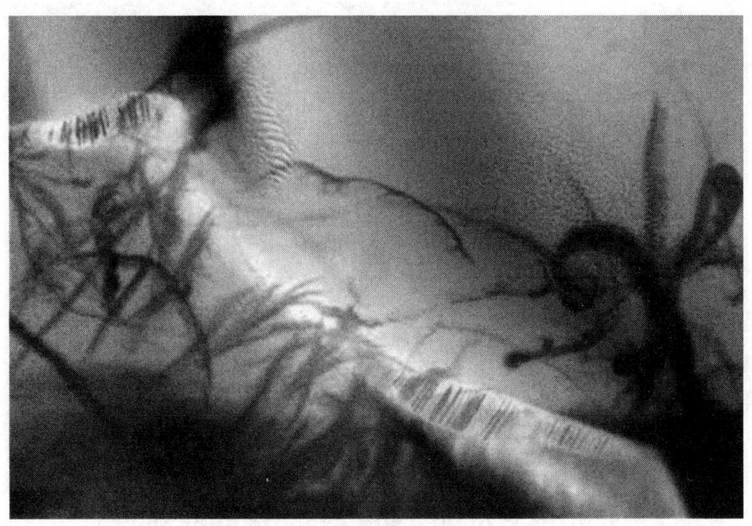

3. "Dust Devils" forming elaborate tracks across sand dunes

Image credit: NASA/JPL/University of Arizona

That was the last real announcement concerning life on Mars that has been officially proclaimed to the world. What we currently have now as the most active scientific program supposedly capable of searching for life on other planets is a flotilla of spacecraft that has been visiting, orbiting and landing on Mars ever since 1998, when the *Mars Global Sur-*

veyor first began science missions and global mapping of the planet.

This is a program of scientific study that has been in continuous operation since then. That's over fourteen years of state-of-the-art, scientific investigation at the cost of several billion dollars. Well, what do we as the human race have to show for that investment? You can certainly say that we've learned a lot about the planet's past and present geology; the amount of water ice on it; its atmosphere and climate. And we've certainly seen some strikingly beautiful and wondrous pictures of the Martian surface and its sometimes strange, alien geological formations.

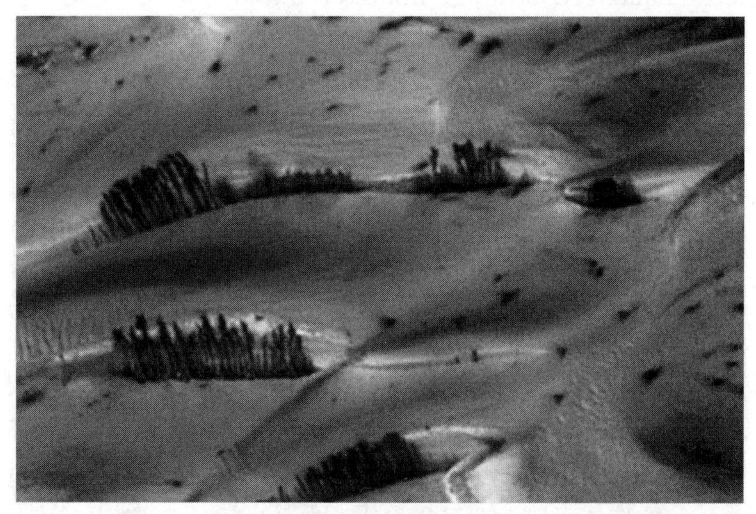

4. Streaks of dust sliding down dune slopes creating the illusion of upright trees
Image credit: NASA/JPL/University of Arizona

But what about the most important question of all? The question as to whether there is *life* on Mars or not? Well, a visit to NASA's website will give you the same official answer as they gave back in 1976 when the Viking landers first attempted to find out if life existed on the Red Planet:

"We don't know yet if there is life on Mars. . ."

And this is where the scientific community should really be scratching their heads and asking some aggressive questions here, because surely, after 14 years of study, NASA must be able to offer something more concrete than "We don't know..."

Consider that scientific data from both NASA and the ESA (European Space Agency) show there are clear requisites for signs of life as we understand it and the conditions to sustain it, present on Mars today:

- Significant amounts of methane in the atmosphere
- Huge amounts of water ice across the planet
- Liquid water consistently appearing on the surface
- Temperature up to 30C/86F at the Mars equator

But NASA will state that Mars is a geologically dead planet. It has no volcanic activity, no magnetic field, only a very thin atmosphere, and the unprotected surface of the planet is bathed in ultraviolet cosmic radiation that makes it very difficult for life to exist there. Yet studies here on Earth, clearly show that 'life' is a very stubborn, tenuous and resourceful commodity and can and does exist in the most severe and inhospitable conditions of extreme heat or cold.

Planetary scientists Janice Bishop and Chris McKay, while investigating carbonate rocks coated with iron oxides collected from the Mojave Desert, found that organisms were protected from deadly ultraviolet light by an iron oxide coating. This suggests a survival mechanism that may have protected life on Mars as conditions deteriorated on the planet's surface.[1]

The Atacama Desert in Chile provides scientists with a useful analogue of Mars due to its hyper-arid, near waterless conditions, one of the driest places on Earth, and here also Dr. McKay found microbial life thriving on the underside of translucent quartz rocks. In this peculiar micro-environment where fog provides the only source of moisture, microorganisms were not only protected against damaging ultraviolet radiation but had enough light, too.

In the same region, Professor Nilton Renno, who is a key investigator for Curiosity, said there is a salt crust on the surface below which exist liquid droplets of salty water that teem with bacterial life.[2]

Life is therefore likely to be found in a wide variety of niches and conditions. It could even exist on an interstellar planet unheated by a sun. Planetary scientists Dorian Abbot and Eric Switzer postulated that a planet ejected from its planetary system could conceivably maintain a liquid ocean under layers of thermally-insulating water ice, deriving its heat and possibly sustaining any life, purely from geothermal flux.[3]

Quite clearly 'life finds a way', and if, as NASA's studies seem to show, Mars, used to be an Earth-like planet with oceans, seas, rivers, an atmosphere, a magnetic field and, very likely under these conditions, 'life', then the chances of that life still existing somewhere despite the cataclysm that struck the planet, must surely be very high indeed.

'Life' will cling on, mutate, evolve, and may even thrive once it has adapted to new conditions, and that is not to mention the 'extremophiles' that—as they do here on Earth—always seem to find some specialized niche and habitat in which to survive and live.

Even if it is found to be true that no life can survive on the *surface* of the planet due to lethal levels of UV radiation, the possibility of finding life *beneath* the surface just a few meters below ground or in sheltered caves far from radiation's harm will have most biologists declare a virtual 99% certainty that life exists on Mars.

And yet NASA will not come out and reflect that view. Instead, they fluff around seemingly waiting forever for that absolute proof. And worse, crucially and infuriatingly, they still won't go *searching* for that proof.

The Mars Science Laboratory: Not-so-Curious?

At the time of writing, the Curiosity Mission in the form of "The Mars Science Laboratory," and at a cost of 2.5 billion dollars, landed successfully on Mars in a hair-raising but brilliantly executed manoeuvre in August 2012. Hailed as the most sophisticated spacecraft to have so far landed on Mars and armed with the most advanced set of scientific experiments yet to explore the planet's surface, what will Curiosity actually do?

Will it be able to search for *life*?

Well, the Viking Mission of 1976 was an astrobiology mission designed to hopefully detect life on the surface of Mars, which, according to some scientists, it did find, and to others, it didn't. Then followed *Pathfinder*, *Spirit*, *Opportunity*, and *Phoenix*, which were all geology missions sent to discover what they already knew—that locations, where the spacecraft landed, were ancient sites where water used to exist millions of years ago.

5. Curiosity: The Mars Science Laboratory (MSL) (Artist's impression)
Image credit: NASA/JPL-Caltech

The Curiosity mission is…well, yet another 'geology' mission!

Well, in reality, quite a bit more than that; a roving, robotic geochemist that can laser zap and powder down rocks and also sample soil to get a real close up look at their secrets, but that's about it. As Project Scientist John Grotzinger said in the NASA video, *The Science of Curiosity: Seeking Signs of Past Habitats on Mars:* [4]

> Curiosity is not a life-detection mission. We're not actually looking for life. We don't have the ability to detect life if it was there.

So, Curiosity was never intended or designed to search for life even though millions of people across the world presumed this was its purpose. How did NASA get away with that one? People heard the media say, "The new Mars spacecraft will be searching for conditions habitable for life," and they presumed, "Oh, they just mean they're going to search for life".

A strange manipulation of public perception was going on there it seems. So, if Curiosity is not looking for life, then what *is* it searching for? Grotzinger continues:

> What we are looking for is the ingredients of life. By that, we mean a place where microorganisms, little tiny single-celled organisms, could have lived and that requires a source of energy and water, because all life as we know it is associated with water. And then we also need a source of carbon.

He explained that the Curiosity rover would be searching for those ingredients, at a mountain in the centre of Gale Crater:

> In that mountain, there's a stack of layers, and like turning the pages in a book we will explore these layers and look at them in terms of whether or not they preserve evidence for ancient habitable environments.

And there you have it. With this spacecraft, they're just looking for habitats that preserve the necessities for life, whether in the past or the present. They're simply asking the question:

"Has Mars ever provided, or does it still provide, a habitat that can support microbial life?"

The fact they already know Mars had a wet, warm past with a much thicker atmosphere would seem to make that question a mite obsolete to my mind. Of course, Mars could have supported microbial life. And yet they are sending a 2.5 billion dollar spacecraft across millions of miles of space just to ask such a pedantic question?

The only important question that needs answering is if life is *still* there and to that end, this should have been an *astrobiology* mission with life-detection equipment, not Curiosity's lame assignment of investigating whether or not conditions have been favourable for microbial life and for preserving clues in the rocks about possible past life.

Well, scientists know that plenty of water exists on Mars and that the sun provides the energy, so all they're looking for now are complex organic carbon compounds and some gases that will betray microbial life activity. If they find these in the rock layers they'll know that there were habitable conditions present for life millions of years ago; if they find them in the soil today then they will have evidence for present-day habitable conditions.

Officially then, the most that this mission can accomplish will be if it can find organic compounds in the Martian soil and activity, perhaps methane gas, that may indicate microbes are present.

Curiosity Will Confirm There is Life on Mars

Discovery of the organics and methane would pretty much confirm that life is present on Mars and therefore NASA *will* be able to declare to the world in a massive display of overblown media ranting that microbial life is alive and well in the Martian soil. The President will then be able to address this marvellous American discovery to the waiting masses. But they will still have to send another spacecraft in a few year's time to actually *detect* the life that exists there.

It is worth noting that Gill Levin and the Viking scientists will be happy because this discovery will confirm that NASA found life back in 1976!

Anyway, can Curiosity discover anything else?

It might. John Grotzinger made a tantalizing remark to the tune of 'we all know water runs downhill'. He seemed to be referring to ancient water in this comment, but what if he wasn't? What does he really hope will be discovered during this mission? A lot more perhaps than the scientific instruments were designed to give him.

There is an instrument onboard that will be able to detect subsurface water up to a meter down but will not be able to reach it. However, if they do manage to find soil that has recently been in touch with liquid water on the surface, analysis of that soil may show signs of present microbial activity.

Curiosity won't, of course, be able to see any microbes, because there is no microscope on board (which makes it a strangely deficient 'laboratory')—but *signs* that life may be present could be revealed. On the other hand, there is the possibility that creatures *larger* than microbes may be seen by the cameras themselves.

I think the most likely discovery, however, in terms of Martian life for this mission, will be *fossils* found in Gale Crater. And that this 'revelation' will be the one hailed to the awaiting world and media.

Get a Webcam to Mars!

In this modern-day and age, it should also be questioned as to why NASA has not sought to engage the public with the inclusion of a webcam and microphone on Curiosity.

Twenty-four-hour video and audio surveillance of the Martian surface would be a really exciting leap in both learning about the real Martian environment and in allowing active participation of the public in Mars exploration and discovery. How many millions would be glued to their monitors in the hope of seeing something move in the Martian soil or hearing the real sound of a live, howling Martian wind?

I suppose sending video and audio data back to Earth must be prohibitive in terms of data radio relay, as even static pictures take hours to transmit I understand. However, where there's a will the inventive and creative imagination of the human mind will eventually provide the solution, so maybe we should just get to it.

Geologists Dominate Mars Exploration

The Curiosity mission, however, is primarily yet another geologically biased science mission as was Spirit, Opportunity, and Phoenix—its astrobiological parameters confined to identifying organics and certain signatures in the soil as possible indicators of life present—no substitute for a proper astrobiology mission. Not since Viking, way back in 1976, has a fully-fledged NASA space mission gone to Mars to search specifically for life.

Between now and then—a span of 36 years—the geologists have dominated the NASA agenda for exploring Mars. While the study of Martian rocks will help us learn how Mars formed and developed as a planet, its value is certainly not worth the enormous costs involved, compared to the monumental discovery and confirmation of actual *life* on Mars.

In my view, it is beyond reason and indefensible to allow for so much money to be expended on such a paltry return of knowledge. And I say paltry intentionally here because despite what interesting data may be exciting planetary scientists, the fact is that nothing we have learned about Mars so far has significantly expanded the awareness and knowledge of the human race in furthering our understanding of ourselves or our place in the universe.

Yes, Curiosity will tell us that millions of years ago Mars had conditions that could have supported life and that there are conditions in the soil today which virtually prove that life is still present. To all intents and purposes, the Curiosity Mission *will* be able to declare to the world that there is life on Mars. However, crucially, the mission is not designed to *detect* that life. So, unless something dramatic is found in the soil—perhaps something moving, or skeletal and fossilized remains—we will have to wait for yet another spacecraft to be launched to confirm a solid, scientific answer to the question, "Is there life on Mars?"

Yet another expensive mission and huge opportunity wasted again. Next up will be the European Space Agency *ExoMars* missions in 2016 and 2018.

But don't hold your breath…the most useful study that *ExoMars 2016* orbiter will be able to offer on the Martian life question (the 'lander' *Schiaparelli*, crashed…), will be to try and identify gases in the atmosphere that may be biological in origin, such as methane, while *ExoMars 2018's* primary objective is to search for well-preserved organic material from the early history of the planet *(yawn, yawn…is that not what 'Curiosity' is meant to be doing? Oh, whatever…)*, although it will have the capacity to drill down for a couple of meters, extract soil and water samples and see what's there, which could be revealing.

Next on line will be the American's in 2020 when NASA will launch a rover mission to search for signs of ancient microbial life (fossilised) in rocks and soil, and to put any promising core samples it finds into caches for later retrieval. The idea being, that one day in the future, the samples will be collected and returned to Earth for analyzes.

(Mmm . . . I'm tempted to predict that won't happen).

The most interesting thing about this rover (Mars 2020) is that it was originally intended to be an *astrobiology mission.*

Yes, it was to have the equipment and experiments necessary to identify and detect microscopic life. Called the *Mars Astrobiology Field Lab Rover (MAFL),*[5] it was supposed to follow the Curiosity Mission as the next logical step in the search for life—but then it was cancelled, supposedly for budgetary reasons.

Yes, I'm saying NASA and the ESA are prevaricating, avoiding, delaying, restricting, and actively suppressing healthy, open-minded, and above all *swift* research that will confirm the existence of life on Mars. But apart from circumstantial evidence that reveals possible limitations that have deliberately been imposed upon present-day Mars missions, can we present anything else that might betray an agenda to restrict the search for life on Mars?

Well, to justify our position we will have to establish a 'pattern of behavior' for NASA on this subject, and the best way to do that is to have a look at the early days of Mars exploration—the mission that remains still the single most deliberate attempt by NASA to try to search for life on Mars—Viking.

The Face on Mars

In 1976 the American space program sent the Viking Mission to the planet Mars with the bold intention of searching for life. They sent two orbiting spacecraft to map and photograph the planet in detail, and two landers to conduct experiments with soil samples.

The experiments were designed to detect the presence of any microorganisms that might be living in the Martian soil, but NASA scientists and authorities decided that the results of the experiments were inconclusive and therefore the mission to search for life, a failed one.

Several scientists still contend that the Viking data supports evidence of microbial activity, even to the extent of circadian rhythms being present,[6] but officially and politically that was the end of the Mars exploration adventure, until 1991 when the Americans sent the *Mars Observer* spacecraft to image the planet in much higher detail than the Viking cameras had been capable of. However, communication terminated with the craft just before it entered orbit and that mission was declared lost.

6. Viking Lander 2 on Mars: Utopia Planitia, 1976
Image credit: NASA/JPL

In between these missions, however, something very interesting had happened back on the Earth. In the summer of 1976, Toby Owen, a member of the imaging team looking for a safe landing site for the approaching Viking Lander 2, and while examining recent photos of the Martian surface, came across one frame in particular, 35A72, which looked remarkably like a human face. It was found in a northern Martian desert called Cydonia Mensae.

Later it was shown to the world's press and presented by Project Scientist Gerry Soffen, as a "Trick of light and shadow".[7] He went on to say that when a picture of it was taken a few hours later it all went away. The truth is, there *was* no picture taken a few hours later, as this whole area was in darkness—it was night. The scientist appeared to have made a crude and dismissive error.

SM1. The Face on Mars: original image (2.5 km) — Viking
Region: Cydonia Mensae, 1976; found by Toby Owen
Image credit: NASA/JPL; digital enhancement: Dr. Mark J. Carlotto

However the journalists knowing no better dutifully accepted what he said, and for three years it was completely forgotten about. That was until 1982, when two imaging engineers, Vincent DiPietro, and Greg Molenaar, began studying the image of 'The Face' wondering if there might be more to it than just a random geological formation that resembled humanoid facial features.

They decided that the existence of a *second* image would provide some basis for analysis, so DiPietro and Molenaar began a search for and eventually found one that had been misfiled. Taken at a different lighting angle it enabled them to begin a scientific evaluation to verify if it was indeed just a trick of light and shadow as NASA had claimed. Their research, however, began to reveal that there was far more to this object than could easily be dismissed as an illusion of nature.

Over a mile long, the striking symmetry of the object with an apparent left and right side, together with what appear to be facial characteristics depicting eye sockets, nose, mouth area and all in the correct proportional and positional relationships, gave serious thought to ponder that this object may well be the product of intelligent design: that some ancient Martian race had created it.

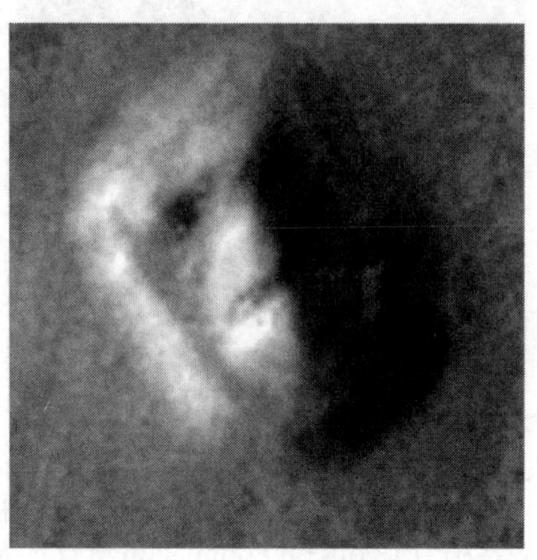

SM2. The Face on Mars: second image — Viking Orbiter
Region: Cydonia Mensae, 1976; found by V. DiPietro/G. Molenaar
Image credit: NASA/JPL; digital enhancement: Dr. Mark J. Carlotto

At this point, you might wonder why NASA themselves had not considered it worthwhile to make a study of this image. Remember that NASA was then and still is the world's leading space agency, studying other planets in our solar system on behalf of humanity. Their integrity and curiosity must be presumed to be intense to deserve this mantle and yet on the single most important potential discovery in our solar system—possible evidence of an intelligent, extraterrestrial presence—they appeared to be not the slightest bit interested.

And this was very strange indeed, especially given the unimaginable breakthroughs possible if indeed alien artifacts were to be found on another planet in our solar system. What could be learned from such a fantastic discovery?

What potential benefits could enhance human knowledge? And on a purely rudimentary level, what would the discovery of an ancient alien monument do for NASA's space exploration budget?

Well, DiPietro and Molenaar went on to find other objects in the Cydonia area with unusual characteristics, in particular, a five-sided mountain that came to be known as the "D&M Pyramid" named after its discoverers.

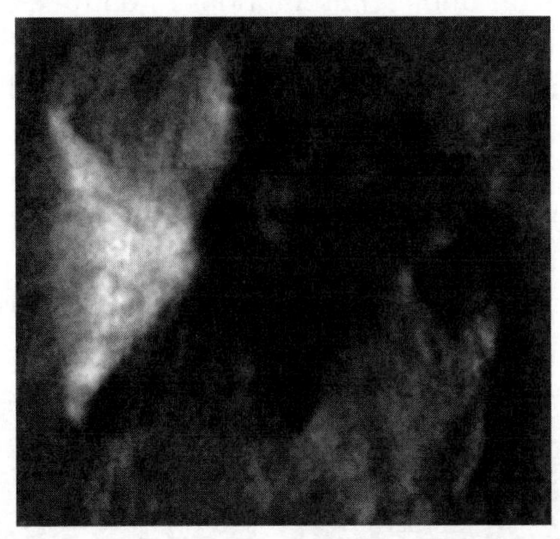

SM3. The D&M Pyramid (3 km)—Viking Orbiter
Region: Cydonia Mensae, 1976; found by V. DiPietro/G. Molenaar
Image credit: NASA/JPL; digital enhancement: Dr. Mark J. Carlotto

Erol Torun, a cartographer and systems analyst with the Defense Mapping Agency, having a degree in geography with a speciality in geomorphology, also conducted a detailed analysis of this object and developed a hypothetical model of its original shape. He concluded that no known geological process could account for the unusual shape and also found that the object is mathematically unique.[8]

These observations confirmed that in addition to The Face we now had a *second* anomalous formation that was suggestive of intentional design and which, perhaps significantly too, lay only a few kilometers southwest of the other formation. DiPietro and Molenaar, along with Dr. John

Brandenburg, later published a booklet detailing this research entitled *Unusual Mars Surface Features*,[9] but NASA once again, at least publicly, took no notice of it.

Former NASA consultant Richard C. Hoagland had also become intrigued with the Cydonian anomalies and the research of DiPietro and Molenaar and began his own investigation which continues to this day. His early research culminated in his book *The Monuments of Mars* published in 1987 along with public presentations to both NASA in 1990 and the United Nations in 1992.[10] The scientific research into Cydonia had now become "The Artificial Origin at Cydonia" hypothesis, AOC for short.

Hoagland analyzed photos of the whole area in which The Face on Mars and the D&M Pyramid were located and also found what seemed to him to be a 'city' of polyhedral objects, including an intriguing trapezoidal formation he later named "The Fort".

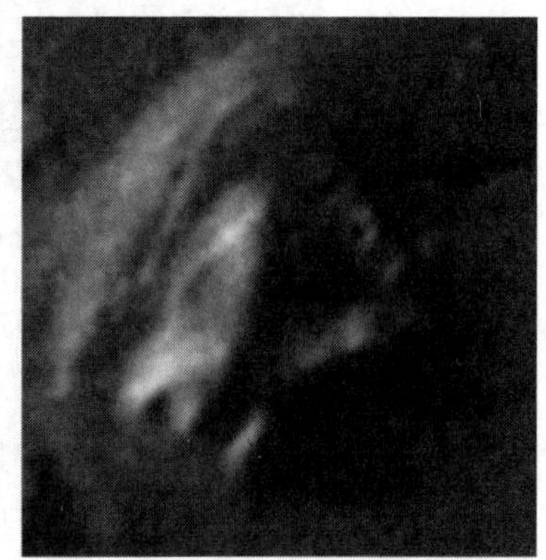

SM4. The 'City' with triangle-shaped 'Fort' (15 km)
Region: Cydonia Mensae, 1976; found by Richard C. Hoagland
Image credit: NASA/JPL; digital enhancement: Dr. Mark J. Carlotto

He also did measurements between the objects and found mathematical relationships that to him began to suggest an intelligent layout. These were developed further by physicist Dr. Horace Crater who found within the complex a high degree of geometric order, statistical analyzes revealing it to be many millions to one against a random chance occurrence in geology.[11]

Dr. Mark Carlotto, an imaging specialist at the Analytical Sciences Corporation, aided researchers by producing state-of-the-art digital image processing of the Viking photos, creating three-dimensional renditions of the Face and the other objects at Cydonia. He also submitted a paper to the planetary science journal *Icarus* detailing his work on producing synthetic views of the Face as if viewed from different perspectives and under different lighting conditions, but his manuscript was rejected on the grounds that it was of no scientific interest! His reconstructions, enhancements, and research of the Cydonia anomalies were summarised in his book *The Martian Enigmas.*[12]

As the case for an 'Artificial Origin at Cydonia' became all the stronger by these growing qualified studies, Hoagland became openly critical of NASA's whole attitude to the subject and their complete lack of interest shown in pursuing the scientific research done by what had by then become a diverse group of independent scientists—imaging specialist Dr. Mark Carlotto, quantum physicist Dr. Horace Crater, plasma physicist Dr. John Brandenburg, anthropologist Dr. Randolfo Pozos, space scientist, and former NASA astronaut Dr. Brian O'Leary, astronomer Dr. Tom Van Flandern, Dr. David Webb, and several other highly qualified scientists in the fields of image processing, geology, geomorphology, mapping, cartography, computer systems analysis, and archaeology.

The issue really came to a head in 1992 as the *Mars Observer* spacecraft approached Mars to begin imaging the Red Planet. To verify and test the research already done by the AOC hypothesis scientists, new highly detailed images were required of the anomalous landforms at Cydonia as the old Viking images were of too low a resolution to be relied upon, and so a campaign began to persuade NASA to target this area.

That public pressure had to be brought upon NASA at all to even consider re-imaging Cydonia is a significant point in itself as it remains inconceivable that NASA did not themselves make it a mission priority. But at every corner and every opportunity, NASA resisted any commitment to use Mars Observer to re-photograph Cydonia or to give any acknowledgment that there was anything of worthwhile note to be investigated at this location. As Mark Pine, Chief of the Policy and Plans Branch of the NASA Office of Space Science and Applications said in a letter:[13]

"NASA has no plans to treat images of the Cydonia region any differently than those of other regions of the Martian surface."

7. Mars Observer: reported lost on approach to Mars, 1993
Image credit: NASA/JPL

Not only that. In an unprecedented move, NASA had arranged that no immediate transmission of Mars Observer photography would be made available to the public. They had arranged a contract with a private company—Malin Space Science Systems (MSSS) who built the Mars Observer camera—that allowed them to sequester and withhold data from the public for up to *six months* while at the same time handing over complete authority to one individual, the company owner and geologist Dr. Michael Malin, as to what images would be taken by the spacecraft.

This clamp-down on public and open access to NASA science data release was suspicious, to say the least, and with Dr. Malin being too an open critic of the Face on Mars and Cydonia research, a clear trend had developed. He gave the following reason for ignoring the AOC hypothesis and landforms when deciding upon the priorities for the Mars Observer Camera:[14]

"The best scientific evaluation available today finds that there is no credible evidence to support the contention that these features are artificial."

A reasonable statement to make, you might think, from a scientist who had presumably been studying a detailed 'scientific evaluation' of these objects. The only problem being that no such report was ever offered as evidence by NASA or its scientists as a way of supporting their view that the landforms are geological and not artificial. Or did NASA carry out an investigation but just did not want to make it public? It's very odd.

It may be telling that Dr. Malin also considered any mission time spent imaging Cydonia and looking for artificial structures to be a waste of time and that the subject irritated him.[15]

NASA's overriding attitude gave the impression that they just thought the whole idea was nothing to do with real science and so they would not give it any mission priority. In short, it was a clear case of prejudice against the possibility that these landforms were artificial. Even the prominent and highly influential space scientist and writer at NASA, the late Carl Sagan, was also dismissive of the research, refusing to acknowledge that the AOC scientists had undertaken any analyzes of note.

However, the blatant lack of scientific objectivity and curiosity shown by NASA had by then drawn the attention of Stanley V. McDaniel, professor emeritus and former chairman, Department of Philosophy, Sonoma State University. He could not contain his disbelief that a respected scientific body such as NASA could be guilty of displaying such a complete disregard for scientific procedure, conduct and discipline, not least with a complete unwillingness to engage with the subject and to duplicate and verify the research already done by qualified, competent and experienced scientists.

In his 1993 book *The McDaniel Report: On the Failure of Executive, Congressional & Scientific Responsibility in Investigating Possible Evidence of Artificial Structures of Mars & in Setting Mission Priorities for NASA's Mars Exploration Program*, he said:[16]

> If NASA's current failure to assign appropriate priority to these landforms remains in effect for future missions to Mars, **science stands at risk of being denied what might be the greatest scientific discovery of all time; and NASA itself at risk of committing the most egregious act of scientific irresponsibility of all time.** Indeed, NASA has already, by its ridicule of the independent investigation and its failures in the area of proper research on the landforms effectively compromised the scientific process. [Emphasis added]

He went on to state:[17]

> I find it extremely difficult to understand what policy does underlie NASA's perplexing behaviour regarding the AOC landforms . . . At the very least **NASA's ongoing behaviour regarding the AOC issue is disingenuous, against the public interest, and demands explanation**. [Emphasis added]

For what is a very high profile scientific organization this behaviour was very unusual, unprofessional, and perhaps worst of all—very unscientific, as Professor McDaniel eloquently makes clear.

One can with reason suppose that there is probably a prevalent attitude amongst mainstream scientists that any talk of 'Aliens and ET' does not belong in serious scientific circles and that the only aliens that should be allowed to be considered are those that were potentially many light-years away on the far end of a radio telescope, way out of harms reach.

And what with the popular media giggling at every opportunity to dismiss extraterrestrial matters with a knowing wink of the eye and a smirking remark about 'Little green men', it could perhaps in this light be understood why NASA's planetary scientists might not allow themselves to become engaged in a matter which might tarnish or damage their scientific reputations amongst their colleagues and employers.

It seems the cards are stacked against ET.

To all intents and purposes then, it could be argued that NASA just played along with this popular tendency to summarily dismiss any talk of E.T. and aliens with a humorous quip. Furthermore, the lack of valid scientific response to the serious studies done by the AOC hypothesis scientists would also suggest that on the matter of extraterrestrial intelligence NASA seemed to automatically shift to a level that took on the form rather of a media and politically motivated organization than a scientific one, effectively relegating the pursuit of responsible, open-minded science and discovery to a back seat.

But is there any precedent or justification for this attitude? Is there any policy or governance perhaps that curtails or restricts NASA's involvement with potential extraterrestrial discoveries and which kicks in as soon as the subject comes up?

Yes, there is.

The Brookings Report

In the early days of NASA, in 1959, they commissioned a report from the prestigious Brookings Institute, entitled *Proposed Studies on the Implications of Peaceful Space Activities for Human Affairs.*[18] This was an important 264-page, year-long study involving a few hundred experts in many fields who were tasked to examine the possible implications of NASA's future space exploration and potential discoveries. Or, in other words, "What possible issues will NASA and the U.S. government have to face when they begin exploring the whole of the solar system?"

Of particular interest, on page 215, was a section titled "Implications of a Discovery of Extraterrestrial Life" and what this report had to say on this subject is striking, although entirely reasonable and rational:

> Cosmologists and astronomers think it very likely there is intelligent life in many other solar systems . . . **artifacts left at some point in time by these life forms might possibly be discovered** through our space activities **on the Moon, Mars, or Venus**. [Emphasis added]

And on page 216 the report makes a suggestion that may throw some light on the peculiar attitude that NASA has shown regarding the anomalies at Cydonia. It reads:

> How might such information, under what circumstances, be presented to **or withheld from the public** for what ends? What might be the role of the discovering scientists and other decision-makers regarding release of the fact of discovery? [Emphasis added]

The report explicitly acknowledged the possibility of finding alien artifacts on other worlds in our solar system and considered the question of concealing information regarding this discovery of signs of intelligent extraterrestrial life. Could it be that, following this report, NASA formulated a policy together with government officials that elaborated on this very point? What to do if extraterrestrial discoveries were made?

But why would they have the concern to hide such information from the public in the first place? Again, the reason may be found in further observations from the report on pages 215 and 225:

Regarding society as a whole:

> Anthropological files contain many examples of **societies sure of their place in the universe, which have disintegrated** when they had to associate with previously unfamiliar societies espousing different ideas and different life ways; others that survived such an experience usually did so by **paying the price of changes in values and attitudes and behaviour**. [Emphasis added]

"Disintegration of society," "Paying the price of changes in values, attitudes, and behaviour,"... have some authorities decided that widespread knowledge of the existence of extraterrestrial, intelligent races will cause human civilization to radically change and perhaps even collapse? If so, would it be a rational reaction to fear knowledge or contact with extraterrestrials?

Perhaps it would be far more likely that a highly advanced E.T. civilization capable of interstellar travel would probably be much more evolved in their sensibilities, morals, and spiritual nature than we are, and that contact therefore with such a civilization may offer us much from which we could learn and benefit from, rather than fear. The report made an interesting observation concerning scientists:

> It has been speculated that, of all groups, **scientists and engineers might be the most devastated by the discovery of relatively superior creatures**, since these professions are most clearly associated with the mastery of nature, rather than with the understanding and expression of man. **Advanced understanding of nature might vitiate all our theories** at the very least. [Emphasis added]

Clearly, quite a revealing document that if NASA had indeed taken it all on board, must have forced officials to sit down and go through their contingency plans with some exactitude as to what they were going to do if they found evidence of intelligent, alien life. Did they fear a meltdown of society and religion back in the 1950s and 60s? That people would panic and wouldn't be able to handle the truth? And most importantly, what do they think *now*?

Is There a Martian Conspiracy?

8. Desolate Mars: why does NASA always present the image of Mars to the public, as a dead, lifeless world? And why do they keep landing their spacecraft at all the boring places?
Image credit: NASA/JPL/Cornell

Most people presume that the sole purpose of NASA is to search for evidence of life in the universe so as to answer the most profound questions on behalf of the human race. But if they have decided not to reveal to us what they find concerning this important question, what is left for them to show us? What do we get from NASA space missions these days?

Well, we've seen it all many, many times now ... endless photos of bleak, desolate Martian and moonscapes; studies of atmospherics and rocks; an apparent search for water; discussion of possible microbial life. . .

Well, now that we have evidence that an official policy may exist that is *restricting* NASA's openness on the subject of extraterrestrial intelligence and its subsequent discovery, this could explain why there is no open research being done into possible signs of intelligent extraterrestrial life on Mars or elsewhere in our solar system. We have what is effectively a closed book on this matter.

But is it fair that NASA should be shouldered with the blame for this situation? Indeed is NASA even in control of this information, or are they, themselves supervised by a more clandestine agency that might deal specifically with the question of ET? If such an agency exists it is certainly not a public one. As Paul Hellyer, a former Canadian Minister of Defence said in a speech in Toronto in 2005 on the subject of UFOs and the reality of extraterrestrial intelligence:[19]

> UFOs are as real as the planes that fly over your head . . . the classification was, from the outset, above Top Secret, so the vast majority of U.S. officials and politicians, let alone a mere allied minister of defence, were never in the loop.

And Dr. Steven Greer from the *Disclosure Project* [20] confirms this statement in rather shocking detail in his article "Understanding UFO Secrecy":[21]

> By the Eisenhower era, the UFO/ET projects were increasingly compartmented away from legal, constitutional chain-of-command oversight and control . . . the

president (and similar leaders in the UK and elsewhere), were increasingly left out of the loop.

Such senior elected and appointed leaders were confronted with (as Eisenhower called it) a sophisticated military-industrial complex with labyrinthine compartmented projects which were more and more out of their control and oversight. From direct witness testimony, we know that Eisenhower, Kennedy, Carter, and Clinton were frustrated by their attempts to penetrate such projects.

This disturbing picture suggests that real agencies are operating above and outside the control of elected office in the United States and across the world and which have direct access and control over extraterrestrial matters. President Eisenhower even warned his nation about this in no uncertain terms in his farewell address to the nation in 1961:[22]

> In the councils of government, we must guard against the acquisition of unwarranted influence, whether sought or unsought, by **the military-industrial complex**. The potential for the **disastrous rise of misplaced power** exists and will persist. [Emphasis added]

And if his fears were well-founded, how might this influence have developed and grown over the last fifty years since that speech was made? We would, of course, be talking about a mega conspiracy theory here. But has a sinister shadow power indeed been developing and growing over the decades? And is there something about the extraterrestrial question at stake that is so valuable, some people have decided they must exert absolute control over it?

Well, in truth, such questions are somewhat beyond the scope of this study, but at least we do now have a workable premise as to why NASA could be resisting the swift release of truth and knowledge concerning the planet Mars and especially any potential discoveries concerning extraterrestrial life that may be found there. It is a premise that states: *NASA may not be in control of the release of information concerning E.T. matters because this subject is classified above top secret.*

OK, it's time to back up all these contentious views I've presented here with some hard evidence. I've been saying that NASA has been dragging its heels on the whole question of searching for life on Mars and that they are simply not doing what should be obvious if they are seriously intent on discovering all there is to know about the Red Planet, and—more importantly—sharing that knowledge with the rest of the human race.

Now to the 'business end' of this book: the images. I'm going to move swiftly and relentlessly through the many official space agency images I'm going to present to you as evidence for:

- Liquid water on Mars
- Vegetation on Mars
- Fossils on Mars
- Intelligent life on Mars

Why am I including 'water' evidence? Well, as every biologist will tell you, liquid water on Earth almost always contains life, therefore the serious search for life on Mars demands that you go aggressively after *any* signs of liquid water that have been discovered. Something that NASA has not been doing.

'Vegetation': little mention yet of this subject and yet if liquid water exists—and they know it does—the possibility of vegetation must be considered.

'Fossils': before microscopic life is ever confirmed by NASA, the discovery of Martian fossils lying on the ground, or embedded in rock is highly likely and in fact may already have occurred and has been kept quiet as some evidence here in this book suggests.

'Intelligent life': this is where things have really gone seriously amiss in the matter of trust that we place in those we presume are out to inform and educate us about discoveries made on our planetary missions. Officially this subject is not even on the scientific radar of NASA, but by the end of this book, I hope that you will have seen enough evidence to be completely convinced that it should be.

Well, you will surely meet with some degree of resistance as to what you are about to see throughout the coming pages. However, I suspect that you may also not be too overly surprised either by what will be revealed to you here. Also, over the years, several images have become popularised as real evidence that Mars has trees, buildings, and living creatures on its surface. I will be doing my best to avoid or dismiss those 'unreal' images for you, in favour of what I believe to be the more serious evidence.

It will, I hope, be both a journey and perhaps a revelation I trust you will welcome.

Chapter 2: Water on Mars

The current scientific assessment of the planet Mars, despite many visits of spacecraft and billions of dollars spent on getting there, remains much the same as it did back in the 1960's—that Mars is a dry, dead, barren, cold planet that is inhospitable to life.

That Mars is a wasteland compared to what it used to be millions of years ago is not in question. However, that the conditions present on the planet today are completely hostile to life I do dispute. And for one reason in particular: the huge amounts of the one vital ingredient that science knows is essential to life as we know it—water.

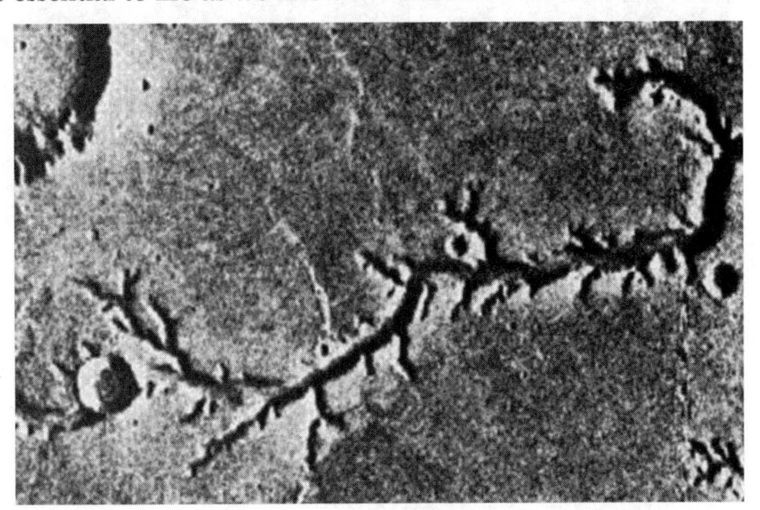

9. Channels carved by water: Nirgal Vallis — Mariner 9
Image credit: NASA/JPL

Water has been known to exist on Mars for a very long time now. Back in 1971, the *Mariner 9* orbiting spacecraft revealed the first real evidence of ancient flowing water having existed on Mars when it photographed river beds, canyons, and signs of water erosion. But also, evidence of water in the present day, in the form of weather fronts and fog.

Then in 1976, the Viking mission made numerous spectacular discoveries showing Mars to have had a long geological history of water existing on the planet: enormous ancient river valleys; deltas; volcanoes that had been exposed to rainfall, and evidence of catastrophic flooding of immense proportions that had left deep scars on the planet's surface.

10. Teardrop-shaped 'islands' in Ares Vallis likely formed by catastrophic floods
Image credit: NASA/JPL

It also found proof that the northern Martian polar cap that had previously been thought to have been made of frozen carbon dioxide was, in fact, water ice, the same as on the Earth. Since then, it has been confirmed that both Martian polar caps consist of 90% water ice, and studies calculate that if this ice were to melt it would form an ocean covering the entire planet up to half a kilometer in depth!

In 2003 the *Gamma-Ray Spectrometer* onboard the orbiting *Mars Odyssey* spacecraft also revealed enormous amounts of water spread over vast areas of Mars, this being a high density of water ice just below the surface. Then, in 2008, NASA scientists also discovered by way of the ground-penetrating radar of the *Mars Reconnaissance Orbiter* huge reservoirs of frozen water on Mars that were far away from the poles, perhaps the remnants of a Martian ice age.

So there is a lot of water on Mars and nearly all of it frozen due to the prevailing subzero temperatures and the low atmospheric pressure. But most importantly, the geology of Mars clearly shows that it was once a world where water in a liquid state flowed for many millions of years, long enough to carve out deep river valleys and canyons, form lakes, and quite probably seas and oceans, too as well.

The key question is, however: was there liquid water on the planet long enough for *complex life* to develop other than microbes? Perhaps plants, insects, and animals…or even more?

Did Mars Develop Complex Life?

Well, if we compare Earth's evolution as a planet, liquid water covered most of the surface for the first 3.5 billion years of its 4 billion year history, with only small expanses of land to be seen. But on this land, there were no plants, insects, or animals. The only life that existed was in the ocean in the form of microscopic life, bacteria, and archaea. Complex life did not evolve until the most recent 500 million years.

If Mars had followed a similar path it too could have had a vast amount of liquid water on its surface at the same time as the Earth did, and yet *not* produced complex life. As astrobiologist Charles Cockell explains:[23]

> Primitive, single-cell organisms had evolved on Earth by 3.5 billion years ago, so something similar might have appeared on Mars. However, it will not have had the chance to evolve as life did in so many ways on Earth.

The consensus opinion is that a major cataclysmic event interrupted the evolution of Mars as a planet capable of nurturing life, and so, even though it is a reasonable assumption that there would have been enough time to evolve at least microbial life, the chances of anything else having lived on Mars are considered slim, as the Mars Science Laboratory Press Pack quips:[24]

"Mars won't have fossils of insects or mastodons; if Mars has had any life forms at all, they were likely microbes."

Quite a definitive statement that, and made with a rather confident assumption. The big question about the Martian past, however, is exactly *when* did the cataclysm strike that caused all of its water to disappear from the surface, and thus determine whether complex life such as plants and animals would ever appear? The timing of this event would appear to be crucial, but it is difficult to pinpoint.

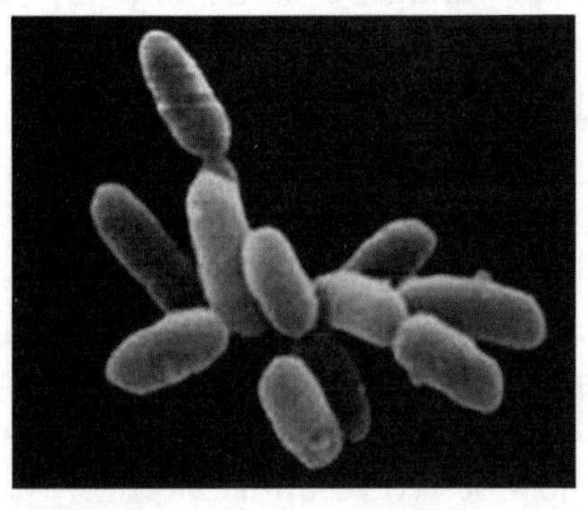

11. Archaea, halobacteria: single-celled microorganisms may have developed on Mars as they did on the Earth
Image credit: NASA

Some scientists believe that Mars gradually changed to the barren world we see today over millions of years through gradual climate change. If this turns out to be true, then any primitive life that existed may have had time to evolve and adapt to the new conditions. However, a catastrophic event such as an enormous meteorite impact would have had a much more drastic effect, causing mass extinction of any existing life forms—and Mars shows the surface scars of having suffered several such massive impacts, although the hardy ability of microorganisms to survive extreme conditions here on Earth will not have passed unnoticed.

Still, given the obvious difficulty in mapping out vast geological expanses of time with any degree of accuracy, especially on another world, and then predicting exactly what conditions may have existed on Mars that may have hampered or perhaps even aided the evolution of life, there seems to me to remain a window of possibility that life of a higher order than microbes, may once have gained a foothold and therefore may still survive today.

But we can understand why NASA scientists are somewhat sceptical towards the idea that anything other than microbes will be found on Mars. They believe that complex life will not have had enough time to evolve in theoretical Martian oceans because Mars became a barren, dead world far too early in its history.

It is a theory and a reasonable one given what we understand of the studies made of the planet so far. But have we studied enough? Have we exhausted all there is to know and learn from the data already received from Mars? And indeed, has enough effort been expended in searching for and exposing data that might throw up a completely different picture for the possible development of life on Mars?

Certainly, the geologists find Mars perplexing. Even the scientists at Malin Space Science Systems, who furnished the camera onboard Mars Global Surveyor, made the admirable admission that they saw processes on Mars they recognized but just did not understand how they came about.[25]

And it is because NASA will entertain no ideas other than a narrowly defined search for primitive microbes that the apparent logic of NASA's scientific quest to search for life on Mars, begins to founder.

NASA Avoids Finding Water

It began to fail for me when NASA's much-flaunted policy of "Follow the water" as a detective's trail approach for discovering any life on Mars, seemed to betray an exactly *opposite* direction to what such a strategy would imply.

—In 1998 the *Pathfinder* mission landed on an ancient flood plain.

—In 2004 the *Spirit* rover was dispatched to where a former lake may have been.

—In 2004 the *Opportunity* rover touched down where hematite was abundant, a mineral known to form where water was once present millions of years ago.

—In 2012 *Curiosity* landed in Gale Crater and is currently trundling around in a place where they know a lake of water used to exist.

Do we see a developing picture? Yes, the real intention seems to be very clear: what they really meant to imply was, that NASA intends to follow all *ancient* signs of water, and avoid all locations which indicate *present-day* water…

The one exception, I suppose, was the *Phoenix* mission in 2008, which at least landed where water ice was present in the frozen soil. But still, liquid water was avoided. NASA may say that for these missions they didn't know of any locations where liquid water was present on Mars, but as we will soon show you, this simply wasn't true because they certainly did know where it was.

Well, there is some hope regarding the latest mission Curiosity, which has touched down in a low-lying area. As the mission's project scientist John Grotzinger said:[26]

"One fascination with Gale is that it's a huge crater sitting in a very low-elevation position on Mars, and we all know that water runs downhill."

Was he just referring to ancient evidence of water activity they expect to find, or possibly more recent signs? Presumably, he just meant ancient water, because the mission planners have made it abundantly clear that this mission, like the others, is strictly focussed yet again on past water activity on Mars where life is not expected to be found. As the Curiosity Press Pack states unequivocally:

"Selection of Curiosity's landing site was not based on traits favoring present-day habitability."

Once again, avoiding a site where they might find signs of life in wet conditions. However, contrary to Curiosity's science objective, the mission's project scientist may nonetheless be hoping to see *some* evidence of present-day water on Mars. Does he know something that we don't, regarding what will be revealed on this mission? Perhaps fate will assist in forcing the hand of those holding up the progress of advancing our knowledge of life on Mars. Maybe the onboard video cameras could yet be a vital tool in exposing some unexpected event if they can be deployed imaginatively and boldly.

Ok, let's get on now and expose further why NASA appears to want to avoid discovering life on Mars.

One scientist explained to me that he believed NASA is not avoiding the search for life but just delaying it because they simply want to avoid the embarrassment of another 'Viking' situation, where they spent two billion dollars to go and search for life on Mars, but could not agree as to whether they actually found it or not. And that as a result of that science failure, they have now adopted a more careful, step-by-step, pedantic approach in which they are now just trying to pick up a 'paper trail' of evidence that will lead them to a location where they will increase their chances of finding life if it exists.

Another opinion argues that NASA's exploration of Mars is no longer confined to the apex goal of discovering whether life exists there or not, but has rather broadened now to include the aim of preparing for a human mission and base on Mars, with eventual colonization of that planet.

Well, be that as it may, we will now see that vigorous and questioning science has not been done, and crucial evidence has been ignored. The fact is, NASA has been avoiding areas where there is known to be liquid water where, theoretically, they could just simply land, dip a scientific toe in the water, look under a microscope and then see if there are any microbes floating about in it, or perhaps even robust Martian tadpoles!

Well, the problem with Mars is that, because of the low atmospheric pressure, which on average is about a 100 times less than the Earth's, any water exposed to the atmosphere will very quickly freeze or boil away into the air, making it surely impossible for any water-based life to survive under such conditions. However, I see no reason why NASA cannot land in these areas and explore the sites where recent water has been active in the Martian soil. An exploration of such locations would surely reveal some activity and remnants of that water's interaction with it, and perchance some microbial activity, or more.

So now to the crux of the matter…

Liquid Water on Mars

Are there any locations on Mars that have present-day liquid water activity? Because if there are, then I believe NASA should boldly be targeting and heading straight for them—not avoiding and landing instead in barren areas with absolutely no water activity at all, as they have been doing.

Well, there is one site they could have explored with their rovers in a region called East Arabia (SM5). This was the first evidence I had ever seen of the existence of liquid water on Mars. The photo was discovered by Richard C. Hoagland during June 2000 while he was looking through the image data released by NASA from the orbiting camera onboard Mars Global Surveyor.[27] NASA had just put up for public viewing on the internet over 20,000 images taken of Mars between September 1997 and August 1999 *(I encourage you to go online and study this image for yourself, as with all the important images in this book. See image index).*

Around this time, MSSS scientists Michael Malin and Ken Edgett had also given a news conference of their discovery of evidence for present-day water activity on Mars as they sought to explain Martian gulley features and dark flows from interior walls and peaks of impact craters. They concluded that these features were best explained by sources of groundwater seepage and surface runoff, proposing a model that described a build-up of groundwater that was kept at bay by a barrier of ice but which periodically burst outwards in a slurry flow of water, ice, and sediment.[28]

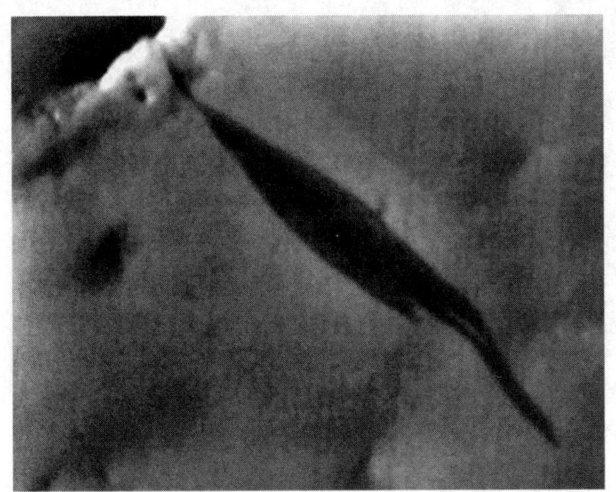

SM5. Water flow from a crater wall (2.5km) — MOC/MGS
Region: East Arabia, Jun 1998; found by Richard C. Hoagland
Image credit: NASA/JPL/MSSS

Other scientists later observed that water flows emanating from isolated peaks and dune crests were likely to be from the melting of near-surface ice or snow originating from the atmosphere during high obliquity—when above-freezing temperatures can occur—or through the process of deliquescence when soil substances absorb water vapor from the air. NASA's generally preferred view, however, was to explain away these flows and stains as 'soil or dust displacements'.

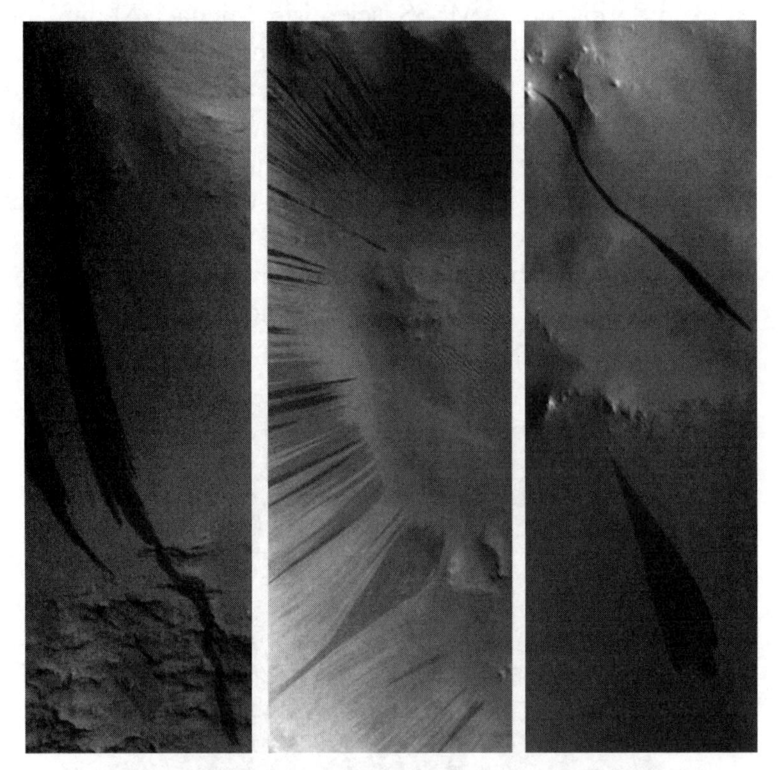

SM6, 7 & 8 Examples of "Dark flows" — MOC/MGS
Image credits: NASA/JPL/MSSS

Although some of these dark flows will be soil slides, many are most certainly not. In the image shown (SM5), the way the dark stain fans and flows outwards surely displays the behaviour of a liquid—and that liquid is most likely to be water.

In support of the water view, researcher Efrain Palermo also points out a clear correlation between the distribution of these flows, falling as they do within the warmer equatorial zone and the distribution of water as revealed by the *Mars Odyssey* spacecraft and its neutron detector and spectrometer instruments.[29] There are hundreds of images of these stains and flows now recorded. Palermo and fellow researcher Jill England have now identified and catalogued many of them, presenting their research to the *National Space Society (NSS)* in Seattle, May 2002.[30]

SM9. "Dark flows", Arabia Terra Region — HiRISE/MRO
Image credit: NASA/JPL/University of Arizona

In the examples shown the liquid originates from a small elevated point and then leaks downwards in a flow that would appear to soak the surface and then leave behind a dark stain. The stains themselves appear to remain for an extended period afterwards, although becoming lighter over time.

As any water in the flow would quickly freeze or boil-off due to the low atmospheric pressure on Mars there may be some residue in the liquid itself that causes the dark staining. Some researchers have suggested the activity of microorganisms. It may however simply be the result of the interaction between the liquid and the soil causing an unknown chemical reaction. As yet the answer is not clear.

What is clear however is that surely there must be a cast-iron case for landing a spacecraft near to one of these flows, to discover something about this liquid. The potential for it harboring some kind of life must surely be worth the journey. At the very least we will be in the act of studying current liquid water activity on Mars, not merely scooping up soil samples from a dried-up ancient floodplain, where water used to flow millions of years ago.

Every biologist will tell you that in almost every environment on Earth where there is liquid water, you will find microbial life. This has to be the guiding principle for the scientific search for life on Mars, so why does NASA seem to be ignoring it?

SM10. Spring & summer flows, Newton Crater — HiRISE/MRO

Image credit: NASA/JPL-Caltech/University of Arizona

Yes, many of these flows are in difficult places to reach—crater walls and steep slopes—but not all of them are. And even so, are you telling me that NASA can't think of some way to reach and extend out some technology to study that soil? If we know there is water there because our images show us there is and our scientists tell us there is, and our scientists also tell us that where we find water we almost al-way*s* find life, then those are the places where we should be landing our spacecraft!

Shane Byrne, assistant professor at the Department of Planetary Sciences, University of Arizona said that a lander would be able to positively identify the existence of liquid water and that such a mission might also be able to hunt for signs of simple life forms.[31]

SM11. Liquid water flow in Gale Crater (15cm) — Curiosity (The rover drove right on by . . . not a whisper from NASA)
Region: Gale Crater, Sol 707, Aug 2014; found by Gary Proffitt
Image credit: NASA/JPL-Caltech/MSSS

Even the Curiosity rover curiously ignored a clear sign of water on its journey to Mount Sharp (SM11). It drove straight past the obvious and unmistakable mark of liquid saturated soil, probably brine. This was a downhill flow of about six inches long that originated from beneath a cluster of rocks, completely ignored and without any public comment from the Curiosity science team. Why did this discovery not make the news in August 2014?

I place this simple question then on the table: if we are truly searching for life on Mars and 'follow the water' is the guiding principle, then why on Earth (or on 'Mars', rather) are we not going to the water?

In this search for liquid water I have begun by showing you some present-day examples of widespread and abundant seepages and flows that could be:

- Water sourced from melting ice or snow that is usually found on high, isolated peaks and dune crests. This water starts to flow when above-freezing temperatures occur during high obliquity.

- Water originating from probable Martian groundwater that appears on the slopes and central peaks of crater walls. Here, the water bursts out from a single source and then flows for a little while down the slope until the source of groundwater empties or freezes.

If it's considered by NASA that these flows may just be soil slides and not liquid water, contrary to what research now indicates, and therefore use this as justification for turning their attention away from these flows in the search for life on Mars, how, I wonder, will they explain their avoidance of the following locations?

Are There Lakes of Water on Mars?

What I am going to present to you now goes far beyond 'seepages and flows' and, according to the current understanding and scientific consensus for the existence of liquid water on the Martian surface, simply cannot happen at all.

And yet, here we have what appears to be *a large area of standing water...*

SM12. A lake of liquid water on Mars? (1.6km) —MOC/MGS

Region: South Polar, Nov 1999; found by J. P. Skipper
Image credit: NASA/JPL/MSSS

Most of the pictures you are about to see were brought to my attention by author J. P. Skipper in his book *Hidden Truth: Water and Life on Mars.*[32]

This one (SM12) shows what certainly looks very much like a flat expanse of liquid water, a lake, or reservoir. The regular, bright edge of the shoreline also suggesting geology that has been eroded and exposed, or minerals deposited as the level of liquid has dropped.

But is this at all possible on Mars, a lake of liquid water?

Here is another intriguing picture that was taken from a fascinating image strip that I suspect will attract some investigation in years to come (SM13). Here we see again the light-reflective strandline perimeters.

But before we go any further we need to point out a massive problem with these pictures...you see they're all taken from *the south polar region of Mars,* where the freezing average temperatures fall between -75° to -120° degrees Celsius (-

103 to -184 Fahrenheit), which is extremely cold. This makes it pretty much impossible for these images to be depicting lakes of pure liquid water!

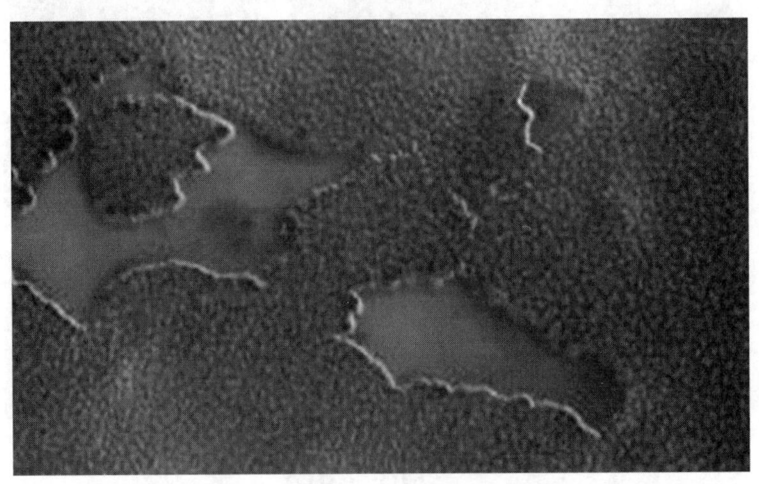

SM13. Lakes of ice or liquid water? (2.6km) — MOC/MGS
Region: South Polar, Nov 1999; found by J. P. Skipper
Image credit: NASA/JPL/MSSS

And of course, it's not only the freezing temperature which surely forbids liquid water but the atmospheric pressure on Mars, too, which is mostly at or below water's triple-point vapor pressure of 6.1 millibars. This means that liquid water on contact with the Martian atmosphere will either *freeze* or *boil away* within seconds of exposure. But not only that: the extremely dry air of Mars would also cause any liquid water to vigorously evaporate.

Logically, therefore, these images cannot possibly be lakes of liquid water, but much more likely to be 'lakes of ice'. For, if liquid water either freezes or boils off in the very dry and low atmospheric pressure on Mars, then it cannot possibly exist in large pools and lakes, as these pictures provocatively suggest. Indeed, you may well ask "Why even bother to look for evidence of standing liquid water on Mars, when the scientific facts forbid its very existence?"

Well, perhaps we don't have all the facts in yet. We are certainly only just beginning to understand the planetary life of Mars and we can be sure there are going to be some enormous surprises to come in the years to follow, as more of this planet's mystery is unveiled. For now, however, we do need an explanation that will allow this impossible evidence of liquid water lakes on Mars to at least be considered, and must therefore simply ask:

Is there any possibility *at all* for water to remain on the surface of Mars in a liquid state, as these NASA images seem to portray? Could there be any conditions in play that might allow this water to remain liquid instead of just freezing, boiling away or evaporating into the atmosphere?

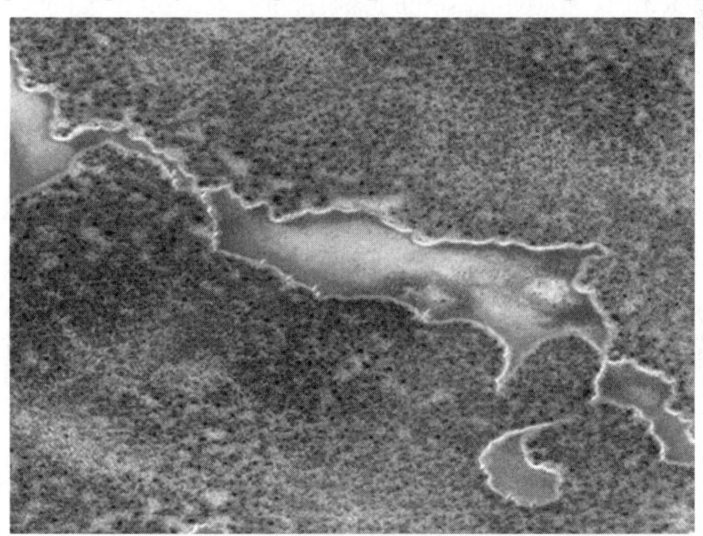

SM14. More lakes? (3.4km) — MOC/MGS
Region: South Polar, Oct 2001; found by J. P. Skipper
Image credit: NASA/JPL/MSSS

In the case of the dark flows, we can surmise that the water simply broke free from the protective, ground-water catchment levels, burst outwards into the harsh Martian air, and then, when the flow subsided, the remaining water just froze or boiled off. However, concerning large pools of standing water, we need an element that will seriously inhibit the process of freezing and evaporation, otherwise, there is just no way that these pictures could be of liquid water.

But on Mars, there is a lot of *salt*, and this could be highly significant. Here on the Earth, saltwater *significantly reduces the temperature at which water will freeze and evaporate*. The higher the salt content the lower the temperature can be before the water turns to ice.

Water with a high salt content is commonly called 'brine', and planetary scientists have concluded that salty water fits the pattern for some of the flows and dark, finger-like streaks we have shown in previous images.

Scientists studying similar flows in Newton Crater and elsewhere in the middle latitudes of the southern hemisphere accept as much, further noting that the flows tend to appear during the late Martian spring and through the summer, fade away in winter and then return the following spring, strongly suggesting that it is frozen, Martian water ice simply melting in the warmer weather.

Alfred McEwen of the University of Arizona and principal investigator for the Mars Reconnaissance Orbiter High-Resolution Imaging Science Experiment (HiRISE), stated:[33]

"The best explanation for these observations so far is the flow of briny water . . . It is more like a syrup maybe, in how it flows."

Strong evidence for the existence of liquid brine came in late 2008 from the Mars Phoenix Lander, when scientists noticed droplets on the spacecraft struts that were growing and moving, strongly indicating they were in a liquid state.

Nilton Renno, co-investigator of the Phoenix Mars Mission, said that he believed liquid water could not exist on Mars because of the freezing temperatures, but after studying the evidence of droplets on the lander, hypothesized that Martian salt in the soil could be keeping some of the planet's surface from freezing, which would allow liquid water to exist.[34]

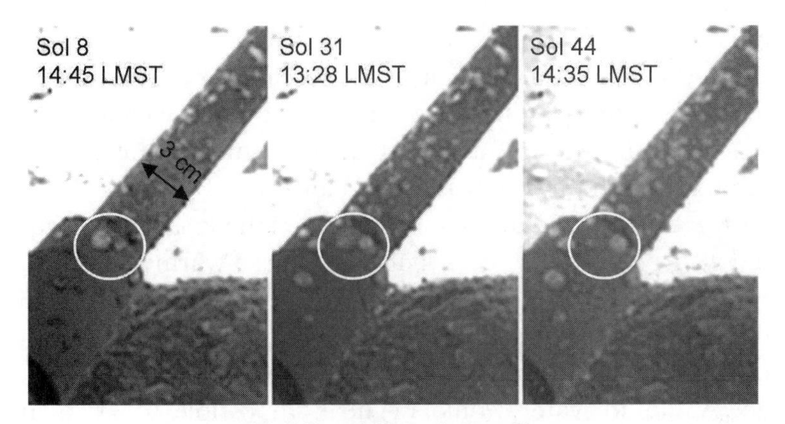

12. Mars Phoenix Lander:
water droplets observed on struts
Image credit: NASA/JPL-Caltech/U. of Arizona/Max Planck Inst.

His hypothesis was later proved correct when analysis of the soil beneath the lander, revealed the presence of salts. Professor Renno explained further:[35]

> We found out that the soil had lots of perchlorate that are a very powerful anti-freeze. We found out we could have liquid saline water at a temperature even lower than we had described.

Brine waters allow the freezing point of water to drop similarly to how when we add salt to melt the snow on roads and footpaths. The perchlorate salts found in the Martian soil around the Phoenix lander, freeze in temperatures of minus 68 degrees C (-90F), to minus 76 degrees C (-105F). The average temperature for the area in which Phoenix landed was -59 *degrees C (-75F)*, which proved that the salts had kept the water from freezing.

So we now have a scientifically credible basis for the existence of liquid saltwater on Mars *[Note: this was 'officially confirmed' by NASA scientists in September 2015]*. But to what extent can that water maintain itself on the surface? Deep within the planet a few kilometers down, extensive groundwater and aquifers were believed to exist across the entire globe, but these have not yet been found. Closer to the surface, scientists feel confined to limit the existence of liquid water to the temporary brine flows we've mentioned which break out on to the surface, and perhaps too, small caches of liquid water gathered a few meters below it.

Overall, the fundamental problem for the existence of liquid water on the surface is that it requires a warmer, thicker, water-laden atmosphere and a higher atmospheric pressure to keep it from evaporating. As on the Earth, pressure decreases the higher up you climb and increases the lower you go. As liquid water would be near impossible to exist on most of the surface of Mars—unless perhaps it was thoroughly saturated with salt and maybe as thick as treacle—that leaves only the lowest-lying areas we can logically consider as potential sites for flowing or standing water, where the atmospheric pressure could just be low enough for any water not to boil away.

Hellas Planitia is at the lowest elevation point on Mars, where the pressure can get as high as 11.55 millibars (0.1675 psi) compared to the planetary average of 6.0 millibars (0.087 psi). In the warmer months, creeks and low-lying valleys in this area could, therefore, be possible candidates for finding liquid standing water on Mars, but even with sufficient atmospheric pressure and a warm enough temperature, there is still the formidable problem of the extremely dry atmosphere which would still cause any water to evaporate very quickly.

So even in the Martian summer, when temperatures at the equator can reach up to 30 degrees C (86F), and with salt-saturated water in low-lying areas being at least possible, the harsh, scientific facts against free-standing, liquid water being able to exist on Mars at all make these images very hard to swallow indeed, especially given that all of the images are located near the deep-frozen, south polar region of Mars. Also, the fact that these images I've shown you so far only originate from that ice-cold region would suggest that these NASA pictures can *only* be of ice. That scientific logic confounds the senses however when we take a look at the next image (SM15).

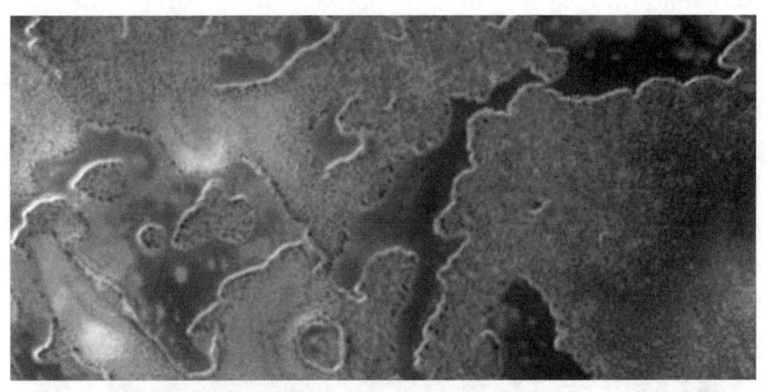

SM15. Group of lakes or polar pits? (3km) — MOC/MGS
Region: South Polar, Oct 2001; found by J. P. Skipper
Image credit: NASA/JPL/MSSS

I find this one truly mesmerizing, and it persuades one to want to believe that it depicts exactly what it looks like: sunlight reflecting off a watery surface. But, if we want to discover the truth then we must always leave healthy room for doubt, and when we are talking about conditions on another planet, even more so. Alien environments are surely going to throw up all manner of strange, wondrous formations and niches; geological, chemical, life. We must be prepared to be completely surprised and most certainly even shocked by what we are going to discover on Mars.

Now, to my eyes, this image appears to be a group of water lakes, a kind of mini-version of the Great Lakes in North America perhaps, but this impression is likely to be 101% utterly and completely wrong. Why? Because we are on Mars, not the Earth. We cannot assume that just because something "Looks like water" that it is so. No matter how much we may want to believe that liquid water lakes exist on Mars, we *have* to do the research.

As such, when I took a closer look at this last image and compared it to examples of what geologist's call "Polar pits"—the result of sublimating carbon dioxide ice, a common natural occurrence in the south polar regions—I realized that this is most likely what we are looking at here, at least in this particular case.

Yes, visually, I would have to say that this picture depicts lots of 'liquid water'. Logically, however, I would have to accept that it is much more likely to be ice, or some unusual, highly reflective geology. All the scientists of course who have studied the scientific data returned by NASA and ESA instruments as to what the planetary conditions are on Mars would have no choice but to declare "Utter nonsense, that can't be liquid water!"

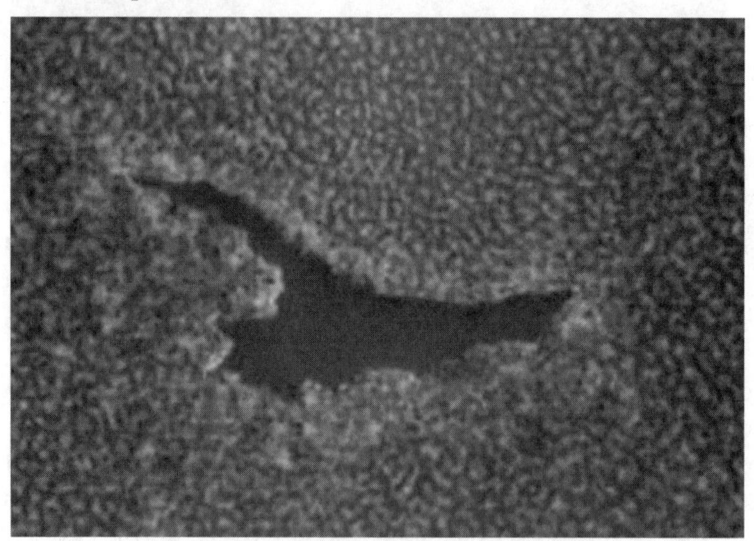

SM16. "Dark lake" (1km) — MOC/MGS
Region: South Polar, Sep 2001; found by J. P. Skipper
Image credit: NASA/JPL/MSSS

Here is another lake-like feature, with a much darker hue to the 'water' (SM16), although this one does not have the appearance of the polar pits as with the previous image. What has NASA to say about such features? Not a lot. I guess they presume the 'dark substance' to be the blue/grey basaltic sand that is commonly seen on Mars and in many cases most certainly is, as the more recent *HiRISE (High-Resolution Imaging Science Experiment)* color imagery proves.

However, some of these images may be suggesting something else, and when we look at them we can but wonder if the science data gathered so far from Mars has somehow missed something.

If you were flying in an aircraft, a few miles above the Earth, and saw this sight below you, what would you presume you were looking at? (SM17)

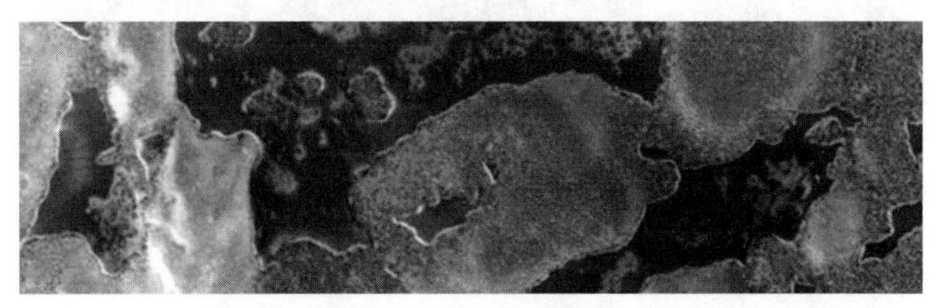

SM17. "Lakes and islets" (10km) — MOC/MGS
Region: South Polar, Sep 2001; found by J. P. Skipper
Image credit: NASA/JPL/MSSS

Yes, I know. Earth-based predilection and such a presumption would be foolish. Unfortunately with the MOC images too, there are none in color, only grayscale. This may be adding fuel for the imagination, particularly as a dark flat expanse seen in a black and white photo instinctively appears like 'water', whereas a photo in color will certainly make a terrain feature, or liquid, much more recognizable.

As yet I haven't come across any HiRISE images of these 'lakes' which might settle the issue, but they'll hopefully show up sooner or later so we can lay this matter to rest. I did ask the tech guys at HiRISE if there was a cross-referencing image data source I could use that would allow me to enter a MOC image number and then see if a corresponding and more detailed, colored HiRISE image available to compare it with, but oddly they said no such database exists.

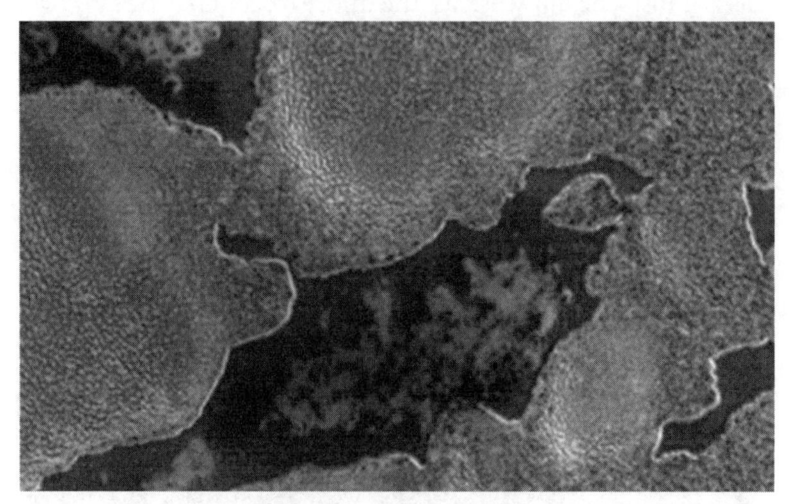

SM17.1 "Lakes and islets"

Once again though, we see a bright strandline surrounding the 'shoreline' that might suggest an even water level is being maintained. But if the analogy with water lake features from Earth is considered foolish, let's take a look at a few pictures from our home planet for comparison.

As we are saying that liquid water on the Martian surface is only possible if it has a high salinity level we may come to find that, if real Martian water lakes do exist, then they may well bear some semblance to *Lake Urmia* on Earth.

Lake Urmia, in Iran, is a shallow body of water only 5 to 6 meters deep that is classified as a hypersaline body due to its extreme salt content. It has what is termed chloride waters—waters that dissolve chloride from rock salt beds.

The most obvious comparison with some of our potential Martian water lakes, is the same bright, exposed strandline sediment around its perimeter, and our notion that any

Martian water is likely to be very salty, as this one is. I should mention that Lake Urmia is over 140 km by 80 km, whereas the features on Mars are a lot smaller, but I think the important thing here is to simply remind ourselves of what a lake of water looks like here on the Earth, and if we are seeing similar-looking geological features on Mars too, then let's ponder that fact.

13. Lake Urmia: Iran, Earth
Image credit: NASA

14. Water lake: South America, Earth
Image credit: ©2013 DigitalGlobe, Google Earth

Here is a familiar shape both on Earth and Mars:

15. Water lake: South America, Earth
Image credit: ©2013 DigitalGlobe, Google Earth

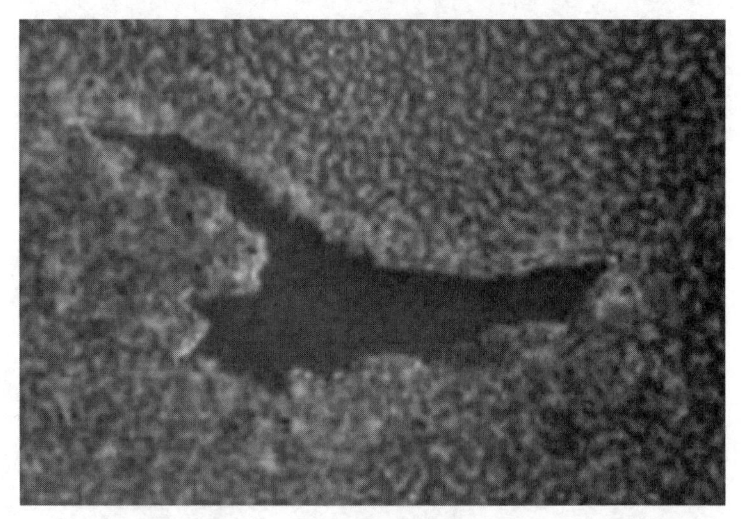

SM16. "Dark lake": South Polar Region, Mars

And to help our eyes a bit instead of flicking back and forth comparing images, let me summarise a few of them to assist the process of data assimilation…

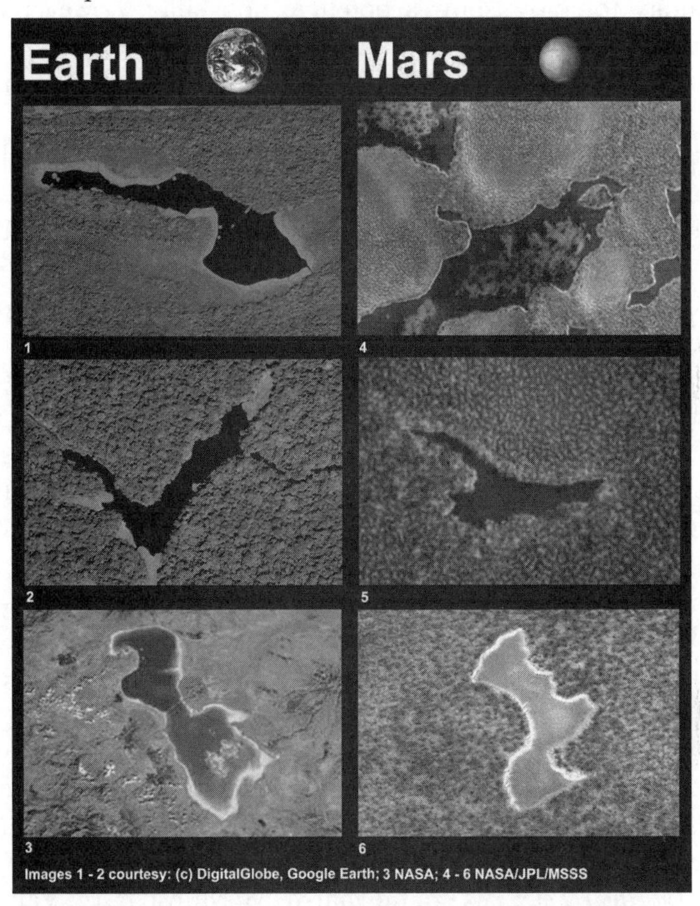

SM18. Earth/Mars comparison: lake formations

Now, forgive me for indulging some artistic license here with these picture/shape comparisons, but I think it's more than justified in helping us to gain a perspective on the evidence presented here. Do we get the sense that water carves and settles into natural, recognizable shapes on both planets?

I believe these pictures from NASA's own MGS spacecraft, which have been dutifully archived online for years now and are available for the public and the science community to study, show a potential possibility, however extreme, that somehow standing water exists on Mars.

Because these examples I've shown *all* exist in the South Polar Region, the only way these lakes could possibly be liquid water to my mind is if they were hypersaline bodies. In fact, so hypersaline they would have to be as thick as treacle not to freeze or evaporate into the atmosphere. But even so, liquid evaporation would still have to occur, which would necessitate a water source being available to constantly feed and replenish these hypothetical lakes.

The deep cold would seem to rule out liquid groundwater, as scientists expect the ground to be deep-frozen to a depth of at least a kilometer and more. However, extensive salt concentrations could be causing unpredicted caches of liquid water to build up beneath the surface or may have resulted in unexpectedly high volumes of water vapor being absorbed from the atmosphere.

Here is a real-life, solid example from our own planet of what we may come to find on Mars, where liquid water exists and indeed remains a liquid even on a predominantly freezing planetary surface. Located on the coldest continent on planet Earth in below-freezing Antarctica, here we have a small body of liquid water *that stays liquid throughout most of the year.*

Don Juan Pond sits in the harsh McMurdo Dry Valleys where it hardly snows, and this very shallow pool of water, usually only a few inches deep, is the saltiest body of water on our planet with 44% salinity. The next saltiest is *Lake Vanda* which is only a few miles away.

Due to its location in a closed basin, scientists see a visual similarity with dried-up basins on Mars. But of much more significance is the process by which Don Juan Pond actually gets its water. For decades it was believed that groundwater, fed this Antarctic pond, but more recent studies strongly suggest that *deliquescence* is the more likely cause.[36]

'Deliquescence' is the process whereby a substance such as salt absorbs moisture from the air until it becomes so saturated that it dissolves and turns into a liquid solution. In the case of Don Juan Pond, this involves salts being deposited on the hills around the pond from incoming moisture from the oceans. This accumulated salt then absorbs water from the air and when it's saturated, trickles down the slopes to feed the basin below. It is the absorption by salts of water vapor from the atmosphere that is believed to be a key explanation for the dark, briny flows on Mars.

Is there any life in this water in Antarctica? Only sparse microflora and bacterial populations have been recorded so far making it perhaps a habitat more suited for extremophiles that can adapt to the hypersalinity.

16. Don Juan Pond: a shallow hypersaline body of water that rarely freezes — does the same exist on Mars? (300m)
Region: Antarctica, Earth; image credit: Samantha Joye

I believe that finding such pools of liquid briny water in low-lying valleys on Mars to be very possible. Whether some of the pictures we have shown are indeed such pools or lakes, remains to be seen.

All in all, we have to face the reality, however, and concede that it is still much more likely that most of these locations on Mars I have shown you were probably once liquid water lakes that have now frozen solid, or, have been filled with dark sand or some other material to give the appearance of water, while those that have a more reflective surface are instead covered by carbon dioxide ice.

Either-or, these areas surely demand much more serious investigation than NASA has given them, which is the whole point of my presenting this material to you.

Now, to summarise this Martian water evidence, there are only 4 possible conclusions to consider.

Either:

1. The images do not show liquid water at all but geology, such as sand.

2. The images do not show liquid water but instead, show frozen water or CO2 ice.

3. The images do show thick, liquid brines, sustained by unrealized local conditions or an as yet unknown scientific rationale.

4. The images do show liquid water, and that is because the Martian science data regarding temperature and atmospheric conditions on Mars is incorrect.

Most scientists will not go beyond explanations 1 and 2 of course, but where do you stand at this point? Well, to close this chapter, let's remind ourselves of the NASA exploration strategy in its search for life on Mars, which is defined by their key phrase "Follow the Water".

Here is a summary from the NASA "Mars Exploration Program Overview":[37]

> Among our discoveries about Mars, one stands out above all others: the possible presence of liquid water on Mars, either in its ancient past or preserved in the sub-

surface today. **Water is key because almost everywhere we find water on Earth, we find life.** If Mars once had liquid water or still does today, it's compelling to ask whether any microscopic life forms could have developed on its surface. Is there any evidence of life in the planet's past? If so, could any of these tiny living creatures still exist today? Imagine how exciting it would be to answer, "Yes!!".

Following the water begins with an understanding of the current environment on Mars. We want to explore observed features like dry riverbeds, ice in the polar caps, and rock types that only form when water is present. **We want to look for hot springs, hydrothermal vents, or subsurface water reserves.** We want to understand if ancient Mars once held a vast ocean in the northern hemisphere as some scientists believe and how Mars may have transitioned from a more watery environment to the dry and dusty climate it has today…**To pursue these goals, all of our future missions will be driven by rigorous scientific questions** that will continuously evolve as we make new discoveries. [Emphasis added]

"Everywhere we find water, we find life"?

"Rigorous, scientific questions"?

Not quite following the promotional message, I would say. My question then to NASA and their Mars Discovery program is this:

"Dear NASA: rather than waste the public's time and money with massively expensive missions to learn about *past* wet conditions on Mars where life *could* have survived, why don't you 'rigorously pursue' your search for life by just landing your spacecraft where you *know* there is water; dip your instruments into a Martian dark stain outflow and see if there are any microbes in it?

"Or, if you're feeling particularly brave, why not land close to one of these fascinating, lake-like features? Some of them are very likely to be made of at least water ice and may even contain organic material."

The Curious Case of Planetary Protection

Well, believe it or not, they say they can't. They're not allowed to. NASA is not permitted to approach any area where there might be a chance of finding *life or liquid water!*

And here's the reason why. . .

On the 28th of September 2015, NASA finally decided to announce to the world that liquid water existed on Mars, and hence the potential for present-day life on the Red Planet had increased dramatically. Journalists and the public leaped in to ask questions:

"Wow, liquid water on Mars! Could there be any water in Gale Crater where the Curiosity rover is? And if NASA finds some, can the rover drive over and check to see if there's any life in the water?"

It turns out there are some dark streaks, RSL (Recurring Slope Lineae) that the rover could visit, but Curiosity is not allowed to go anywhere near the water yet as Jim Green,[38] NASA's Director of Planetary Science explains:

> It's not as simple as driving a rover to a potential site and taking a scoop of soil . . . Not only are these on steep slopes, we need to ensure that planetary protection concerns are met. In other words, how can we search for evidence of life without contaminating the sites with bugs from Earth?

So now they know where to find some liquid water they can examine, is NASA trying to come up with an excuse not to investigate it for signs of life? They are saying there is a risk that robotic equipment may contaminate Martian water with microbes from Earth, and if that happens, then scientists will never be sure that they have found Martian life, as it could have come from our own planet. Which in itself is a reasonable and sensible precaution to take and indeed that is not really in question. What *is* in question is why this issue of 'contamination' has only recently come to the fore, at least with regards to public awareness of it.

Let's remind ourselves again of NASA's ongoing mantra that has been trumpeted as a fanfare for Mars exploration over the last decade or so:

"Follow the water! Water is key! Everywhere we find water on Earth, we find life!"

Yes, amazing isn't it. Despite NASA's assertion that they want to specifically look for "Hot springs, hydrothermal vents or subsurface water reserves", they are now stating that if they actually *find* some water, they are not allowed to even touch it! Now how does that work for heaven's sake? This suggests to me at the very least, that not much planning or thought has gone into what NASA was actually going to *do,* once they found the water they were searching for (if indeed they were honestly looking for it).

So far during this book, we have argued that despite landing three rovers on Mars with an ongoing mission to search for water as a precursor to finding life, NASA has somehow managed to avoid all the known liquid water sites. Although the protection of Mars as a unique scientific opportunity to search for life on another planet is a commonsense reason for being very careful and to take sensible precautions regarding contamination, what does *not* make sense, is NASA claiming to the world that they are searching for life, when in fact they are most definitely NOT doing so.

As Lee Billings, an editor at *Scientific American* succinctly states:[39]

"In the quest for extant life on Mars NASA has been judiciously avoiding the very places where it may most likely be found."

NASA has known for over forty years exactly what they need to do if they want to search for life on another planet as it is not a new problem. Remember that in 1976 they sent Viking to Mars, which had to be 'oven-baked' before it was allowed to land.

17. A Viking Lander...about to be baked at oven temperature so it can search for life on Mars. No spacecraft has been sterilized to this standard for planetary protection since 1976
Image credit: NASA/JPL

Also in 1967, the U.S. signed the United Nations Outer Space Treaty which stipulates how planetary bodies need to be protected during our activities in space, *expressly forbidding* harmful contamination of other moons and planets by Earth's biology.

There is also another organization called COSPAR (International Council for Science's Committee on Space Research) that sets out the rigid protocols for space-exploring nations to follow, including what categories of *sterilization* are required for spacecraft before they are allowed to land on any planetary surface. For Mars and the search for life, COSPAR category IV has to be adhered to, with further subcategories that determine just how much sterilization is required for the craft. This is an especially important matter to understand if we want to grasp the reality of what's involved in the search for life on Mars:

Category IVa—Landers that do not directly search for Martian life (Opportunity and Spirit were classified as such).

Category IVb—Landers that search for Martian life but which do not come into contact with any liquid water (Curiosity*)*. These have to be sterilized to 300 bacillus spores per square meter and 300,000 for the entire spacecraft, the same as category IVa.

Category IVc—Landers that search for life in a "Special Region". On Mars, these are areas where liquid water is known to be present and where the temperature is sufficiently warm enough that scientists believe Martian life may exist, and also where Earthly organisms may propagate too. Only the Viking landers have ever been sterilized to meet IVc standard, which involved being sterilized by dry heat up to 125° Celsius (257F) to restrict microbial populations to 30 spores for the whole craft.

So of course, NASA knows all about these restrictions, for they have to discuss the sterilization requirements of every new mission with the planetary protection officials of COSPAR. This has to happen even before the craft is designed, as parts and materials have to be chosen that can withstand the sterilization procedures required.

Now we know why Spirit, Opportunity, and Curiosity had to deliberately avoid areas where life might exist. They weren't sterilized enough to go anyway near them.

But the real issue surrounds these "Special Regions" on Mars, the sites where it has been agreed that Martian life may theoretically exist because of the likely prevalence of liquid water and warmer climes. And crucially, where scientists believe that microbes from Earth could also possibly get a foothold and thrive if they successfully jumped off the lander onto 'damp Martian soil'.

So how do they decide exactly what locations on Mars are to be deemed Special Regions that must not be contaminated with Earthly organisms?

Well, as Mars is examined ever more closely each year and new studies and scientific assessments are made of the planet's potential for harboring and sustaining life, COSPAR is updated with these evolving results by the most up-to-date science of the day . . . provided by NASA . . . who then duly recommends new areas on the planet that should be designated Special Regions.

NASA tells COSPAR which areas on Mars they should not go to because there might be life there.

Now, this may be viewed as somewhat suspicious and indeed counter-productive, especially if you are trying to tell the world that you are searching for life but then make deliberate arrangements to avoid finding it! However, it also makes sense insofar as NASA is the agency sending most of the spacecraft to Mars, and their scientists are the ones doing all the learning about Mars. So really, who else is better positioned to update COSPAR on what should be designated a Special region? But we might also ask, just how convenient an arrangement that would be if you were trying to avoid finding life as well.

So, what's wrong with this picture? Well, I believe it can be nailed down to the *sequence of missions* sent to Mars. It seems to me that if their goal is to truly search for and discover life, then the missions they are sending and the limited scientific objectives built into those missions, are just not following a logical path.

Unless I've missed something glaringly obvious, the mission sequence for discovering life on Mars is a straightforward two-step procedure that should go something like this:

Step 1: NASA maps out "Special Regions" on Mars. These are areas where potential Martian life may be discovered (Task completed).

Step 2: NASA sends an astrobiology mission to one of the Special Regions to search for that life. The rover is sterilized to IVc standard to reduce the risk of contamination by Earthly organisms (Task yet to be done).

So simple! But we don't get that. What we get instead is this:

Step 1: NASA *avoids* the Special Regions on Mars. These are areas where potential Martian life may be discovered.

Step 2: NASA instead sends the Curiosity mission to a "*Non-special" region* to search for habitable conditions for ancient life. The rover is sterilized to IVb standard, prohibiting it from entering areas where present-day life may be found.

As we can see, they are avoiding what they should logically be doing. They know where liquid water is to be found. They know the best regions in which to search for present-day life. But they WON'T send a mission to Mars to look for and detect that life.

And if we still need persuading that this is actually what is going on, then we only have to consider the fact that NASA *still* has no current plans to progress to a definitive astrobiological search for present-day life, because their next billion-dollar rover mission due up for launch is *Mars 2020,* which will *(surprise, surprise...):*

- Search for signs of **ancient microbial life**.
- In a '**Non-special'** region.
- It will be sterilized to **IVb standard**.

It's the same with the rover mission due from the European Space Agency (ESA), *ExoMars 2020,* which will also be sterilized to the same standard and will also not be landing in a Special Region. This despite the claim that one of their scientific goals for this 'search for life' mission is to:

"Reach a location possessing **high exobiology interest** for past or **present life** signatures."[40] [Emphasis added]

Well, we all know where they should be sending their rovers now if they are seriously interested in discovering life on Mars, but they won't go there. This means we can deduce the following facts:

- They won't sterilize their spacecraft to the degree necessary to enter the Special Regions and so begin a search for present-day life.
- They choose instead to continue a search for ancient signs of microbial life that lived millions of years ago in Non-Special regions.
- The so-called "Special Regions" where life is most likely to be found are clearly very much off-limits.

In short: **NASA is <u>NOT</u> searching for present-day life on Mars**.

Planetary scientists are nonplussed and frustrated at this avoidance of the next, very obvious step to take:[41]

Chris McKay: "It seems pointless to send missions to search for life into non-Special regions if the Special regions are the regions that are of interest for possible life."

Nilton Renno: "If we're looking for life, we should really go where life is most likely to be."

And the situation has become even more ridiculous now that water has been found at Gale Crater where the Curiosity rover is roaming about.

Gale Crater was supposedly chosen as a landing site because they did not expect to find liquid water there in any significant amount, or habitable conditions either that might contain present-day life on Mars, or provide a new home for earthly microbes that may have hitched a ride on Curiosity.

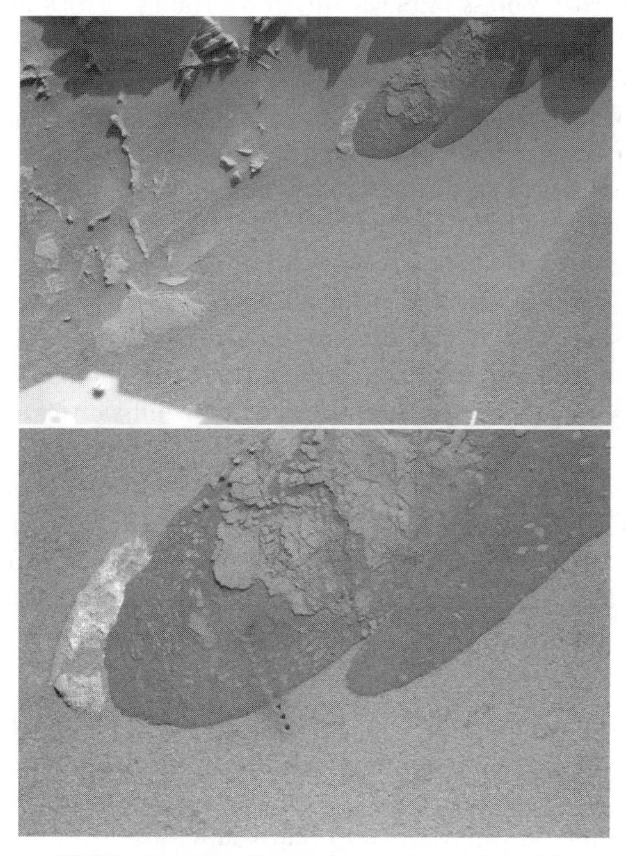

SM19. Possible frozen water is analyzed by Curiosity's laser (10cm)
Region: Gale Crater, Sol 1349-50 May 2016
Image credit: NASA/JPL-Caltech/MSSS

In fact, the NASA Planetary Protection Officer Catherine Conley even helped choose the landing location herself. However, when a colleague showed her an image of the likely presence of water, from Sol 707 (SM11), she became alarmed and instructed the Curiosity team to keep an eye out for further signs to avoid possible contamination.[42]

With NASA having no choice now but to specifically watch out for signs of water, they inevitably came across a small frozen mass underneath an overhang of rock and decided to target it with the onboard laser to get a read-out of its chemical composition (SM19).

Unfortunately, we don't know the results yet as the data is being sequestered for six months (!!!), but at least this development has finally got NASA screaming and kicking to focus on present-day water.

As the rover climbs Mount Sharp, it is coming closer to some dark streaks of RSL that have been observed on some slopes by the HiRISE camera aboard the Mars Reconnaissance Orbiter. Two of these RSL are close enough for Curiosity to reach, pending approval of a mission extension. The stated science goal would be to get images of these potential water sites from a distance of a few miles away and see if there are any seasonal changes and to also confirm if they are indeed briny flows and not dry avalanches. Of course, what people really want to know, is if there is any life in the water, but Conley reiterated her problem of maintaining the planetary protection protocols:[43]

> Kilometers away—it's unlikely that it would be an issue. In terms of coming much closer, we need to understand well in advance the potential for Earth organisms to come off the rover, and that will tell us how far away the rover should stay.

So what are they going to do? It is said there is some heated debate going on concerning planetary protection protocols and the obvious scientific necessity to study this water. Let's hope the scientists who seriously want to look for life on Mars win the debate.

Once again, what I believe we have here is obfuscation and prevarication surrounding a subject that will reveal world-changing knowledge and discovery. We shall see over the coming months how this plays out. My expectation, however, is that they will likely try to keep the science restricted to just analyzing the 'amazing discovery of liquid water' on Mars and confirming and re-confirming that it is indeed water and drag that out for as long as possible.

Incidentally, the ChemCam drill onboard Curiosity could potentially have examined recently water-soaked Martian soil for signs of life, but before Curiosity was launched a deliberate decision was made to open the box containing the drill bits in a less-than-sterile environment.[44] When the Planetary Protection Officer heard about this, she had to reclassify the mission to forbid Curiosity from touching any ice or water. NASA could have made a historic discovery, but someone ensured that it could not happen.

That is one helluva missed opportunity: going all the way to Mars at a cost of 2.5 billion dollars, and then not being able to check out some water that you find there. I wonder who made that decision.

So let's sum up. We will have at least two more future landers on Mars who will not be able to declare there is life on the Red Planet. They will possibly find signs of life in the Martian soil, but because the scientific instruments won't be sterilized to the IVc standard, there will always be the shout of, "We can't be sure—there is the possibility of contamination by Earthly organisms. . ."

It is quite unbelievable and indeed may I suggest even quite farcical, that for years the public has been duped into believing NASA is searching for life on Mars, while the whole time they were deliberately avoiding the areas where life might be found. But there we have it. We will draw upon our infinite reserves of patience. In the meantime, however, we can continue to work out what they are really up to.

Will Astronauts Land on Mars before Life's Found?

So, for the next five years at least, NASA will have no astrobiology mission on Mars that can actively search for and detect life . . . and the years roll quietly on by . . . however, they do plan to land *astronauts* on the Martian surface by the mid-2030s, a prospect that seems to interest them a whole lot more than finding life. Could it be that NASA is planning to just get to Mars and set up base before life is discovered, and to begin a colonization process? Is this their real priority?

But I wonder how such a vision will play out, especially with the private companies also vying for a stake in Mars such as Elon Musk's SpaceX and the major defense contractor Lockheed Martin. NASA is using the word 'cooperation' a lot when speaking of the private sector and their growing role in funding aspects of the space program, so maybe we should keep an eye on that.

President Obama said in 2010 that he expected to be landing astronauts on Mars during the 2030s and NASA Administrator Charles Bolden has reiterated this aim. More recently Ellen Stofan, the NASA Chief Scientist and advisor to Bolden made a very clear and unambiguous statement recently as to how she thinks the question of life on Mars will need to be answered when she said:[45]

"I strongly believe we will never settle this question of determining whether or not there's life on Mars unless we get human scientists down onto the surface of the Red Planet."

John Grunsfeld, former astronaut and NASA Associate Administrator for Science, also stressed how imperative it was that we . . .[46]

"...Send astrobiologists and planetary scientists to Mars to explore the question of, 'Is there current life on Mars?'..."

It would seem then that NASA is now making its heel-dragging science program of searching for life on Mars, co-dependent on their human exploration plans for colonizing the planet. A policy that was openly demonstrated to the public when we saw the Ridley Scott movie *The Martian*, blatantly promoted by the space agency.

I mean, can we honestly believe that there was no coincidence involved when the premiere of *The Martian* came only four days before they announced liquid water on Mars? And then two weeks later it was released to movie theatres in America?

A film, incidentally, which never mentions at all the question of whether there is life on the Red Planet. The focus of the story is simply on the science involved in how to establish a base on Mars and stay alive. No coincidence there, I'm sure.

Does NASA have an urgent need to get to Mars before life is discovered? Is this what is really behind all these geological, non-astrobiological missions that are deliberately not equipped to detect life?

As we have seen, the space agency continues to delay and avoid the chance of discovering life and now appears intent on pushing the discovery back even further to the 2030s, when it hopes to be landing astronauts there. Therefore, I won't be holding my breath for an announcement before then.

Well, if NASA won't go looking for Martian life for real, let's continue with our own investigation and see what we can dig up for ourselves.

I believe that we have presented enough evidence to stir up some serious questions regarding the existence of liquid water on Mars and NASA's reluctance to investigate it. Now we are going to turn up the heat a bit further and ask if there are any signs of what we usually find on the Earth whenever water is present—vegetation.

Chapter 3: Signs of Life on Mars

In this chapter, we won't be discussing the search for microbes and microscopic life forms on Mars, because there is plenty of information on that subject elsewhere. The purpose of this book is to focus on what NASA is *not* searching for, and to that end, we will now take a look at some possible evidence for plant life.

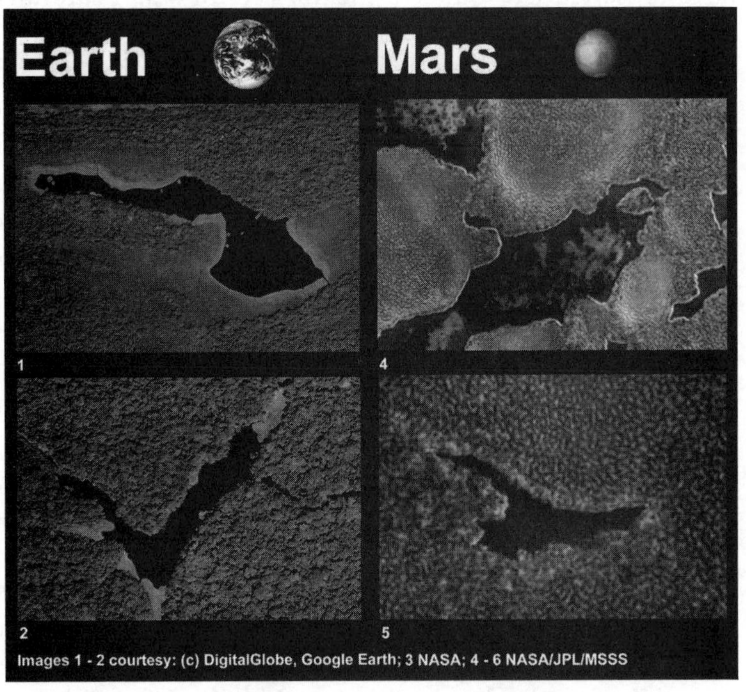

SM18. Earth/Mars comparison: lake formations

Do Martian Trees Exist?

If we take a look again at the comparison images on our chart (SM18), you will see that two of the Earth lakes clearly show an abundance of *forest growth* surrounding them. But what is it that surrounds these possible Martian lakes? We would presume it is something geological—rocky terrain, polar defrosting features, etc. But if these lakes are water—

and we must guardedly maintain the position of 'if'—it would surely be a possibility that they could be nourishing Martian *vegetation* on their shores. This happens on the Earth around water and so, logically, it could be happening on Mars.

Such bodies of water however would need to be replenished by a water source, and as there are no surface rivers or rainfall on Mars (although it does snow!) the lakes would have to be fed by aquifer's, melting ice, or water-saturated salts through deliquescence.

With extensive water ice embedded in the Martian soil pretty much across the entire planet and plenty of salt to lower its freezing point, it could be surmised that if any plant managed to take root and draw upon subsurface water, as well as from the lakes themselves, then vegetation and organisms might just be able to survive and perhaps even proliferate in these locations.

Notwithstanding the severe ultraviolet radiation, the predominantly freezing temperatures, and the extremely dry atmosphere, could Martian plant life indeed have evolved to adapt?

As one planetary scientist declared to me, "Once life has started, it would be hard to eradicate entirely." Would it be possible therefore that at the time when Mars' lakes and rivers were not frozen but flowed freely for millions of years, such vegetation as would have hugged the shoreline and fed from those waters may still survive today in some form, because it had evolved and adapted to the harsh conditions of present-day Mars?

- Could tough, hardy plants be flourishing now on the Martian surface today?
- Could the knobbly terrain surrounding the surely frozen lakes possibly be a tough variety of scrub and bush?

As we saw before with the Martian water question, the NASA images of the Martian surface can be both revealing and deceiving. We will, therefore, proceed with an open-minded but cautious air, as we now consider what evidence might exist for vegetation and trees on Mars.

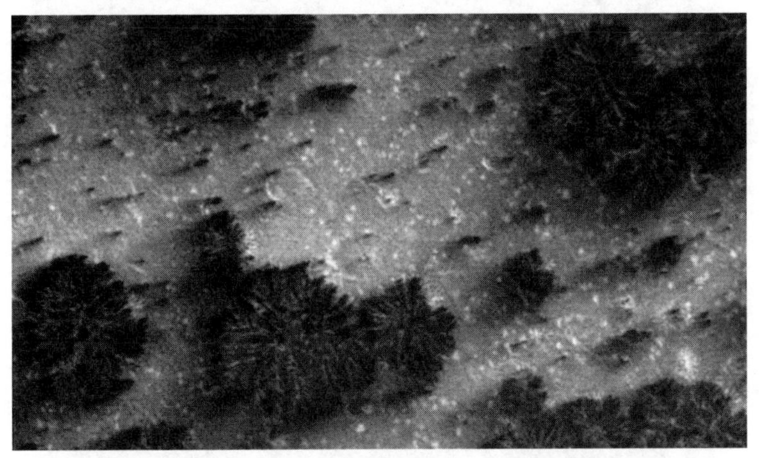

SM20. Trees on Mars? NASA description: "De-frosting South Polar terrain" (6km) — MOC/MGS
Region: South Polar, Oct 1999; image credit: NASA/JPL/MSSS

This is a section of a much longer image strip taken by the *Mars Global Surveyor Orbiter* and released to the world in October 2000 (SM20). If you haven't seen this image before—and many will have done—you would be justified in letting out a gasp of surprise, for indeed this looks amazing!

This picture has since become popularised as proof that trees and forests exist on Mars and it's very easy to see why. I too was taken aback and excited when I first saw this image some years ago.

Now, the resolution of this picture, unfortunately, won't let us squeeze much more detail out of it, but close examination of this photo will at least show what appear to be unmistakable tree-like branches spreading outwards from a central core in a very recognizable terrestrial pattern. I think the similarity in this image to living, growing trees is overwhelming. And here's something fascinating: the *size* of these things . . . the largest formation in this picture is about *half a mile across!*

SM20.1 Trees on Mars? (1.63km)

If these are indeed living, growing plants, then they must be drawing on extensive subsurface water and nutrient supplies to be able to grow to such immense proportions, as we don't see any surface evidence of water in the vicinity. And we are talking about a whole forest of these things, too.

Upon viewing the above images, even the great sci-fi author, science writer, and futurist, Arthur C. Clarke, was moved to comment. He said in a radio broadcast in 2001:[47]

"I'm quite serious when I say have a really good look at these new Mars images. Something is actually moving and changing with the seasons, that suggests at least, vegetation."

He said that these particular formations looked to him something like "Banyan trees".

NASA scientists explain them as some kind of bizarre geology; a cyclic freezing and defrosting phenomenon. Such an explanation would, of course, bear the hallmark of the scientific consensus view reflecting the limitations for Mars as a habitat for life, and under such restriction, a geological explanation will always take precedence over a 'life' explanation. However, in this case, I believe they are right.

We have seen some truly bizarre forms on Mars that almost defy scientific explanation, as many present-day Martian environmental conditions are beyond our Earthly experience. And too, some of the actual images returned by the Mars Global Surveyor orbiting camera are not all they seem to be either.

Apart from optical and interpretative issues, such as 'inversion' (where a surface feature that is cut *into* the ground can be visually mistaken for a feature that seems to be *raised above* the ground), there are a few images I've seen that simply do not look right. They are over or under-processed or something, and on occasion, it is enough to make me wary and cautious of what I'm really looking at.

But this image of potential life is so important that I feel we need to dig deeper. You see there is just something that bothers me about this image of 'Martian trees' because it seems too good to be true. Here's why:

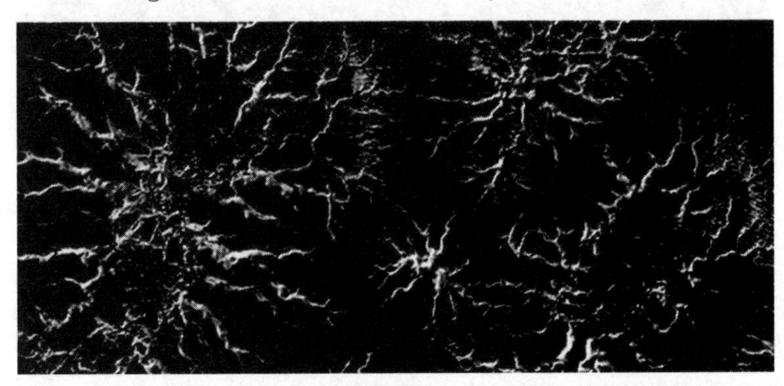

SM21. "Martian Spiders" — HiRISE/MRO
Region: South Polar, Jan 2011
Image credit: NASA/JPL-Caltech/University of Arizona

The preceding image that was taken by the superior HiRISE camera onboard Mars Odyssey (which is capable of much more detail than Mars Global Surveyor), shows what's called "Araneiform terrain": spider-like formations that are a widespread feature in the south polar landscape of Mars during the springtime (SM21).

The current explanation for them begins with the action of sunlight that warms the surface below a layer of carbon dioxide ice that has formed during the winter. The ice then turns to gas, which, as it expands and is now under pressure, looks for a weak point—a crack from which to escape—which it does so, bursting out onto the surface as geysers, carving channels and spewing dust. The carbon dioxide gas then freezes and settles back into the cracks again to form this bizarre pattern.

18. Spring begins in the Martian South Polar icecap: jets of CO2 gas laden with sand, burst out from underneath the surface to shoot up into the sky
Image credit: Arizona State University/Ron Miller

I found this next image while searching for another picture of the 'giant trees', to verify the nature of the branch-like structure (SM22). This one again was taken by the HiRISE camera onboard and shows terrain which looks remarkably similar to the image of our 'forest of trees'. I wonder if this area was specifically targeted to settle the question.

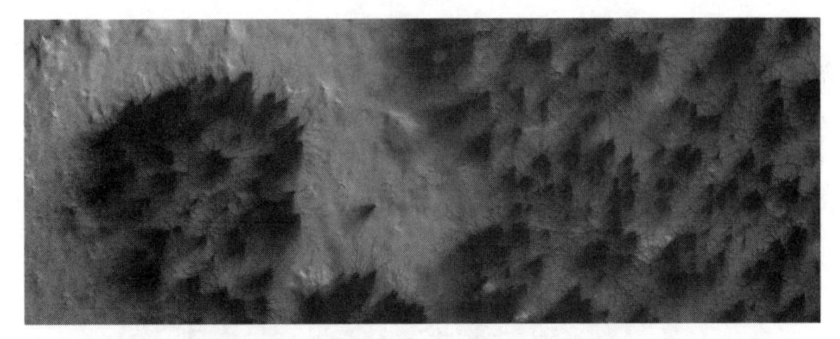

SM22. "Starburst Spiders" (2.8km) — HiRISE/MRO
Region: South Polar, Feb 2009
Image credit: NASA/JPL-Caltech/University of Arizona

The radial, branch-like pattern is revealed clearly here, and when we compare it with our 'trees' image again, I think what is clarified is the illusion of 'dark shadows beneath the branches'.

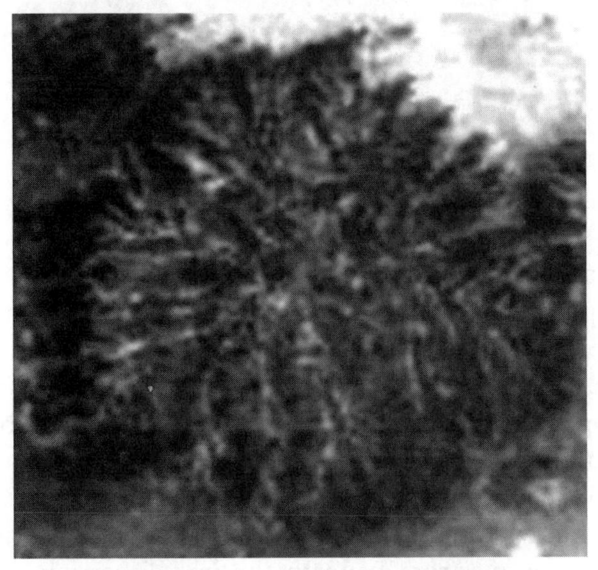

SM20.2 Trees on Mars? (750m)

The lesser detailed MGS image (SM20.2) gives the impression of huge trees and branches casting shadows beneath them, but a careful look at that image will show that the dark patches are not at all consistent with shadows formed by the direction of sunlight, but much more closely match the nature of the dark patches in the HiRISE image, which are formed by *dust trails* left by geysers venting carbon dioxide gas as shown in the close-up and enhancement (SM22.1).

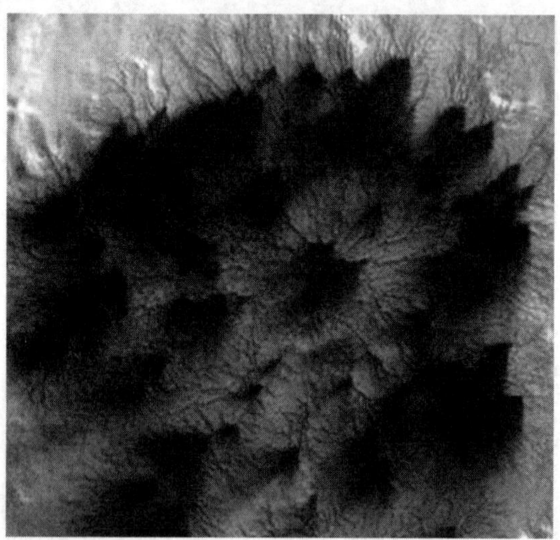

SM22.1 "Starburst Spider" (750m)

Perhaps there is something in the image resolution of the MGS picture, the lighting and overall quality that lays this picture open for confusion and misinterpretation. Or, maybe, the spider formations grow, expand and change to such an exotic degree that they may even look a bit like trees for a while, as they do here.

The images collectively could also be displaying the simple cyclic change of the spiders. They begin to appear when the CO2 gas erupts from beneath the ice, dispersing soil and creating branch-like channels in the ground. Then the gas freezes and falls into the cracks leaving highly reflective, white channels that give the illusion of ridges. The CO2 ice later sublimates and disappears, leaving behind the empty channels in the ground that have been colloquially termed the "Starburst Spider" pattern we see (SM23).

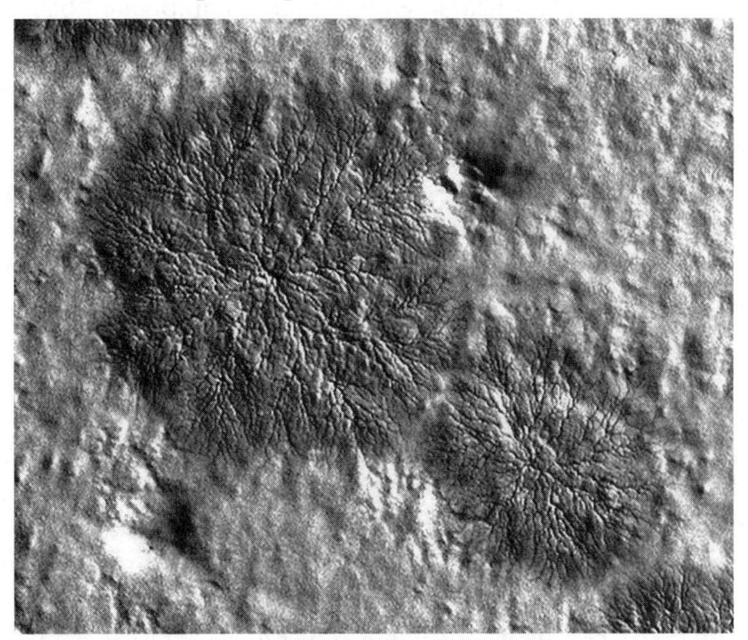

SM23. Starburst Spiders in the springtime: CO2 ice has turned to gas, leaving behind cracks in the surface — HiRISE/MRO

Region: South Polar, July 2011
Image credit: NASA/JPL-Caltech/University of Arizona

Are NASA Images Being Sanitized?

There is another line of thought that casts suspicion over the fact that the MGS images show the features as clearly 'ridged' and 'upward', appearing tree-like, whereas the more recent HiRISE images show the features as more ground level-based, with 'channels' cut into the ground, suggesting perhaps a deliberate attempt at inverting the images so that they don't appear like trees anymore. Well, as this book is suggesting that evidence of life on Mars is at worst being obscured and suppressed, and at best is just avoided, it would not be a surprise to this author that image manipulation is being undertaken by those who want to control the release of this information.

Are images from Mars being digitally manipulated before they are passed onto the science community and the public for analysis? If such a deplorable situation were indeed in effect, the ramifications would be enormous and would throw the whole world of science into turmoil.

Well, disturbingly, there is evidence that this kind of thing has taken place in the past. In one particular case, it involved doctored images of the Moon, where anomalous objects had been airbrushed out of pictures for public consumption *(see chapter 5)*, so it's something we can't rule out.

In particular, that evidence was based on actual photographic images, whereas everything today and the vast majority of the images presented in this book are all from a digital computerised source. Once images are converted to data streams and numbers a much greater scope for manipulation is inherently offered, so if suspicious agencies were so inclined, I would have to say that it is certainly within the realms of possibility that image data transmitted from Mars could be passed through a 'sanitization filter' that could target, hide or disguise anomalous objects before they arrived for analysis by science teams.

Bearing in mind what we are suggesting in this book—that discoveries made of extraterrestrial signs of intelligence are deliberately being kept hidden—such a program of information blackout, suppression, and manipulation, would of necessity be in operation. But does that mean we should discount every image we see from Mars as being possibly suspect?

No, not at all, there are so many thousands of images of Mars now that no human being could ever look through them all. This means that any manipulator would have no choice but to rely on a *computer program* to go through them and to recognize, isolate, and hopefully mask them from public view with some kind of 'anomaly identification' algorithm.

They would also have to be careful not to raise suspicion and from what we've seen so far in the images, clearly, quite a few anomalies remain. Reliance on technology, together with a fearful predisposition, will always miss something, but more likely I think, only the glaringly obvious will be targeted for 'editing', while the majority of strange things that by the nature of this subject will be open to interpretation to some degree, will be left untouched for people to argue over.

So, are these Martian trees or not? I'd go with the planetary science geologists on this one. As much as I would love to see evidence of trees on Mars, I don't think this is it. The fact that these widely observed Martian spider features display the exact same radial, branch-like patterns as our 'forest of trees' and also given the general problem with imagery resolutions and such issues that we have to consider in our equation too, means we should avoid this particular image as evidence of plant life on Mars, and conclude that what we are looking at is what has been termed "Araneiform terrain": spider-shaped formations caused by the pressurized venting of CO2 gas from beneath ice sheets during the Martian spring.

They are not trees. I do think, however, that in this case, NASA could have done the public more of a service by putting this information out more robustly. They must have known that this image of "Martian trees" was doing its rounds on the internet for years, but as far as I know, they did nothing about settling the issue clearly and effectively so that the public could learn all the facts about it.

Above all, it is the truth we want to know about. The truth about what's on Mars. And although some will dismiss official explanations just because they are 'official', the fact is that most of the time NASA's highly skilled, highly specialized, experienced scientists have a pretty good shot at things and know what they're talking about.

Therefore, if we want a real examination of the reality on Mars, my position is that we have to take the established science on board for the most part and then see what they may have perhaps missed out, or more pointedly, where they have decided they won't go, for whatever reasons.

Martian Vegetation and Plant Life

So, now that we have dismissed the most popular mythical photo concerning trees on Mars, are there any more contenders that might reveal possible plant life on the Red Planet? Here's an interesting one where we could be forgiven for imagining this photo as depicting a picturesque alpine scene of snow and conifer trees from Earth (SM24)! Close examination of this image shows an array of objects, small and large, emerging from the ground and extending vertically as trees would do, and perhaps the suggestion also of surrounding areas of lower-lying 'bushes and scrub'.

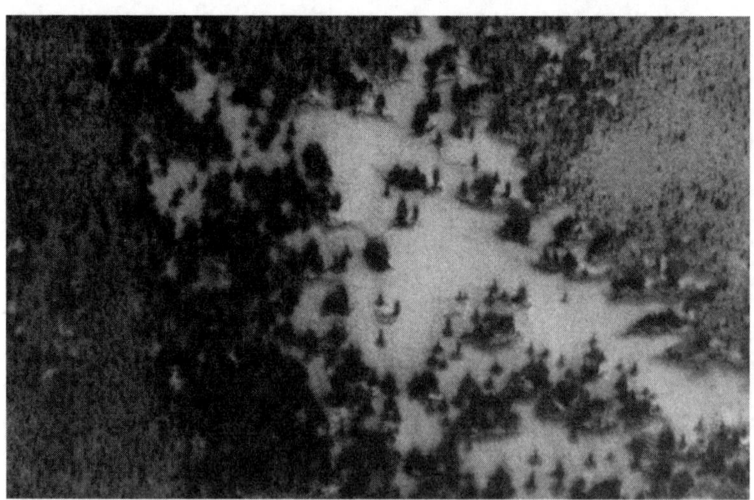

SM24. "Alpine scenery" (4km) — MOC/MGS
Region: South Polar, Aug 2001
Image credit: NASA/JPL/MSSS

SM24.1 "Alpine scenery" (detail)

We can also see root-like tendrils creeping along the ground from the base of the 'trees'—which is probably a sign that we are again just looking at the usual araneiform terrain associated with this south polar region. Although it may be wise not to discount the possibility that some of these 'tree root' structures may be exactly as they appear.

SM25. "Martian bushes" (4.5km) — MOC/MGS
Region: South Polar, May 2005; found by J. P. Skipper
Image credit: NASA/JPL/MSSS

Here's another very interesting image (SM25), released to the public in 2006 and officially described again by the MGS imaging team as "Defrosting South Polar terrain" (terrain where 'spiders' are widespread). This is just a small section from a much larger swathe of this area. How the dark areas spread and clump together is a familiar pattern that reminds us of terrestrial bushes and vegetation. It seems to look less like a geological pattern but more so a biological one.

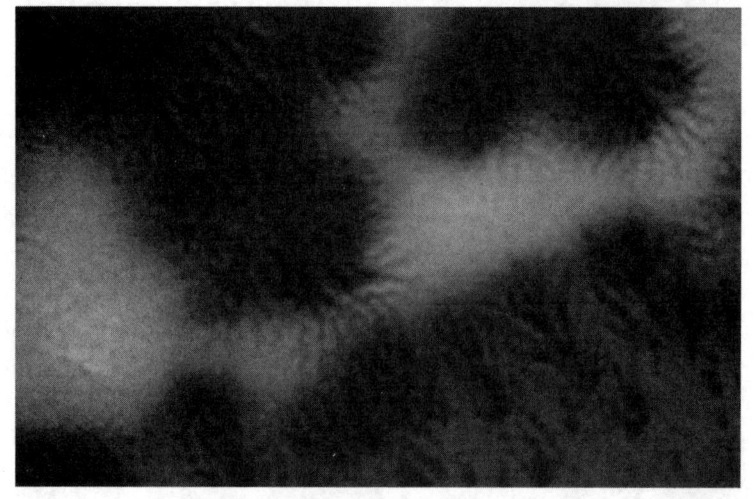

SM25.1 "Martian bushes" (1km)

Here, in a close-up view (SM25.1), we can see branch-like tendrils creeping out from the 'bushes' and extending outwards over the lighter surface, reminiscent of a growing life form. However, now that we are aware that this pattern is caused by venting CO2 gas—the spiders—we have to keep this firmly in mind when observing potential vegetation growth like this, and also allow for a large variation in what bizarre formations we may see that result from the Martian polar regions defrosting in the springtime.

Arthur C. Clarke mentioned the seasonal movements observed on Mars, the dark swathes that increase and expand from the polar regions during the warmer months and then diminish when the cold returns, which for many years many believed hinted at possible vegetation coming and going as subsurface water thawed from the ice, creating conditions habitable for life.

Given the appearance of spiders during this time of the year, and the consequential CO2 geysers spewing out dust to create the 'moving dark areas', there is now little talk of vegetation as an explanation for these observations seen on the Martian surface, but have they thrown the baby out with the bathwater?

For, if the sunlight is warming the ground beneath the layer of carbon dioxide ice and turning it back into a gas, as is the case with the formation of spiders, could the sun also be defrosting the *water ice* there too, and in so doing, creating a habitat for potential or dormant life to re-awaken cyclically? Could we, therefore, have both scenarios in play and perhaps also some resultant life growth in the same vicinity? It would surely be a wise thing indeed not to exclude the possibility of vegetation living in some life-sustaining niche on Mars, where water ice often turns to liquid and the sun provides all the light and warmth necessary.

In the next example (SM26) we have indications of defrosting in the hexagonal pattern featured in the light areas at the foot of the picture, but what is this dark grill-shaped material extending downwards towards it?

A close-up view reveals a very interesting texture to the parallel, tubular forms, certainly looking very organic, with a strong semblance to hedgerows here on Earth (SM26.1). The spiky, branch-like protrusions could, of course, be a crystalline or rock structure, or a bizarre and exotic phenomenon that has built up over millions of years, uninterrupted and free to develop into whatever Martian nature has to offer.

SM26. NASA: "Defrosting dark spot" (3km) — MOC/MGS
Region: South Polar, Oct 2010; found by J. P. Skipper
Image credit: NASA/JPL/MSSS

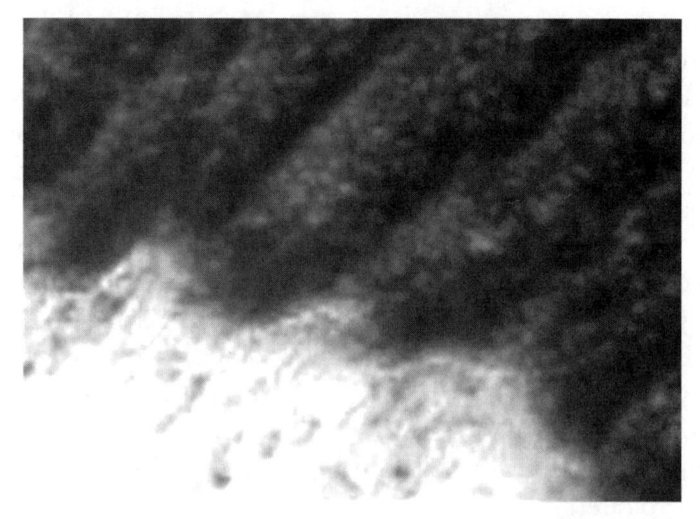

SM26.1. Detail of dark "Hedgerows" (1.4km)

Whether it is life or geology, I cannot reason why NASA would find a site like this one so easy to avoid.

Apart from the difficulty in trying to identify possible evidence for vegetation on Mars from pictures that are not quite detailed enough, there is another problem-area and one to which I have alluded to earlier—that of 'optical illusions'. Studying orbital imagery can and often will confuse eyes!

As we have discussed already with the 'trees', what can appear to be something extending *over and above* the surface is just as likely to be a formation *cut into* the surface. It is a trick of the eye caused by inversion where you can see something in opposite relief to what it portrays in reality.

This problem is compounded by 'flipping' where the orientation of the image is changed from its original orbital viewpoint. This can mean a flip from north-south, south-north, or east-west, west-east, confusing the visual data for the viewer. One would presume that the images uploaded to NASA websites for public and scientific study, will always display the correct alignment, however, this does not always seem to be the case.

On balance, therefore, I would have to say that images which show potential evidence for vegetation are fraught with the possibility of misinterpretation of geology, simply due in part to the optical difficulty involved in correctly assessing them, and I believe should therefore not be assessed as serious evidence for now until we can get much better pictures. That is not to say that I discount these images as possible evidence of vegetation, I am just not convinced enough yet that they are, based on what I'm seeing.

The next image falls into the same category (SM27). It could be tantalizing evidence of vegetative, biological growth crawling over the Martian surface or another example of exotic Martian geology. This one is particularly interesting because of the twisted fern tree pattern, inviting us to believe that we are seeing a thick, tangled, interwoven mass of vegetation—which, incidentally stretches for many miles around. However, upon studying the surrounding areas of this location on the full original MSSS image the light coloring and optical conditions persuade me that we are probably looking at geology.

SM27. Martian vegetation or geology? (700 m) — MOC/MGS

Region: South Polar, June 2005; found by J. P. Skipper
Image credit: NASA/JPL/MSSS

The next strange image I want to show you appears to display some kind of growth pattern, reminding one of the bacterial spores that grow on a laboratory petri dish (SM28). The conventional explanation for this one is that we are looking at a carbon dioxide ice-mass sat on a pedestal crater and that the dark spots are simply the soil being slowly revealed from beneath, as the ice gradually sublimates and disappears. Until we can land and take a closer look at this kind of strange feature we will have to accept generalized explanations. But I wonder. It's the circular shapes that intrigue me about this picture.

Skipper had an interesting idea to explain them.[48] He suggested this is a water-ice mass, not CO2, and that this crater is fed by an underground liquid water upwelling upon which the ice mass sits. More so that this situation has given rise to a habitat in which micro-organisms appear, living in the multiple, circular, dark-edged shapes we can see, which he says may be shallow pools of liquid meltwater.

SM28. Martian "Petri dish" (3.7km) — MOC/MGS
Region: South high latitude, Oct 1999; found by J. P. Skipper
Image credit: NASA/JPL/MSSS

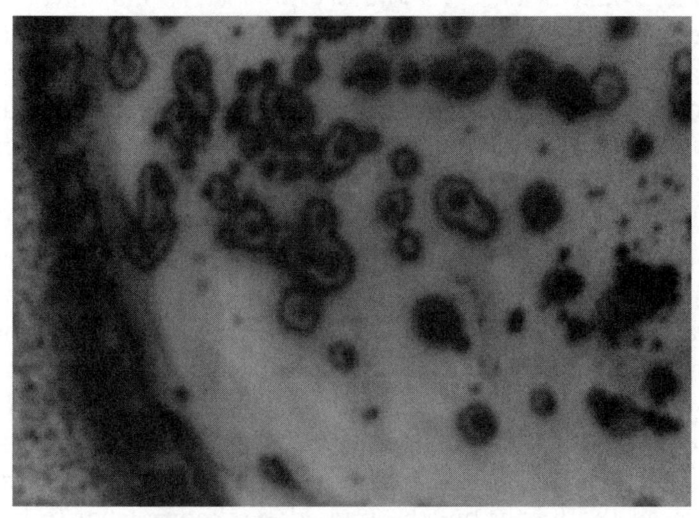

SM28.1 Martian "Petri dish" (900m)

With the abundance of water on Mars, at some stage, microbial life will have started using that water to proliferate, and sooner or later I believe we are going to see evidence of that activity. Whether this picture itself reveals such life I cannot say, but I do consider it worthwhile to broaden our expectations of what we might come to discover, instead of purely cataloging every unusual feature on Mars as 'defrosting CO2 ice'.

I would expect this prevailing attitude to change once microbial life is finally accepted as existing on Mars. The scientists will then start to view unusual features with a more discerning and inquisitive eye, actively searching for further signs of life. Until that new paradigm is accepted however, that Mars is a proven home to life, then scientists will unlikely venture beyond their peer-reviewed comfort zone. But surely, yet again, what another fascinating location this would be for a NASA spacecraft to land and discover who-knows-what.

Martian Fossil Evidence

The Mars Science Laboratory prospectus unequivocally states that we will not discover any fossils of animals on Mars. How they can be so sure I find surprising, given that the planet once had oceans, seas, lakes, and rivers, and therefore a realistic chance of evolving life beyond the microscopic kind.

There was plenty of water, so mud, sediment, and conditions to preserve fossils of creatures must have been very similar to that of the Earth back then. Therefore I think it is irrational for scientists not to consider the possibility of finding some remains of that potential life while their exploration rovers Spirit, Opportunity and now Curiosity trundle over the Martian landscape.

Distinguishing a fossil from a rock is a difficult task even here on the Earth, but the trained eye knows what to look for. I wonder if any palaeontologists are studying the Mars images for signs of fossils. Given that NASA does not believe there will be any fossils to find I suppose they have not assigned anyone to look for them. In the same way that they are publicly not employing archaeologists to study images for possible signs of past civilizations. They have a closed mind on these subjects that should be of serious scientific interest. It is for them to answer why.

SM29. Martian skull? (12cm) — Spirit Rover
Region: Gusev Crater, Sol 513 Jun 2005
Found by J. P. Skipper/E Lucena; image credit: NASA/JPL/Cornell

Well, here are a few pictures that might stir them out of their complacency. This humanoid, skull-like object, is most probably just an odd-shaped rock (SM29), but is NASA's confidence that no sign of animal or intelligent life will be found on Mars based on open-minded science or prejudice? Are they prepared for a surprise?

Here is an image that has been discussed as possible evidence for what could be the skeletal remains of a small rodent-sized creature (SM30):

SM30. Animal skeleton? (9cm) — Spirit Rover
Region: Gusev Crater, Sol 823 Apr 2006
Image credit: NASA/JPL/Cornell

The shape and curvature of this object are indeed suggestive of the spinal column of an animal, but we should remain mindful concerning the nature of the rocks in this location. A perusal of the original rover image and the geological nature of the landscape will reveal many porous-textured volcanic rocks in the vicinity, some partially buried in the sand such as this one (19):

19. Porous volcanic rock — Spirit Rover
Region: Gusev Crater, Sol 853; image credit: NASA/JPL/Cornell

The random chance that one such rock might resemble 'skeletal remains' is therefore very high, and this evidence should be assessed as such. It is, however, an intriguing object that one should keep an open mind on, I feel.

Let's now look at a sample of the Martian surface with a varied collection of small stones and fragments scattered about (SM31). It appears to be like ten thousand other images from Mars that show a myriad of rocks of infinite shape and variety, either partially buried in the soil or fully exposed.

So is there anything unusual about this very common looking scene? Something we can identify as being out-of-place?

SM31. Fossil field? (0.5m) — Spirit Rover
Region: Gusev Crater, Sol 016 Jan 2004; found by J. P. Skipper
Source image credit: NASA/JPL/Cornell; Graphics: SecretMars

At first glance, maybe not, but what happens when we zoom in closer to this picture? Several interesting pieces now begin to pop up that may be worth examining:

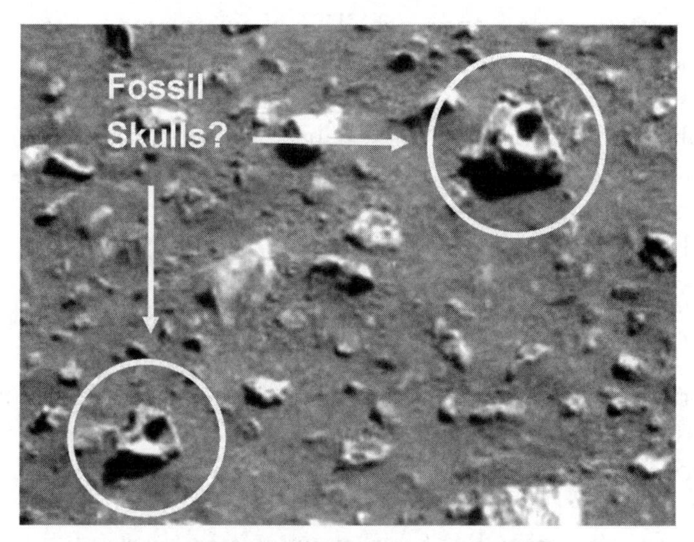

SM31.1 Fossil field? Objects 1 (1cm) & 2 (2.5cm)

The above examples have been presented by some researchers as evidence of fossil skulls of small creatures (SM31.1). Let's zoom in some more:

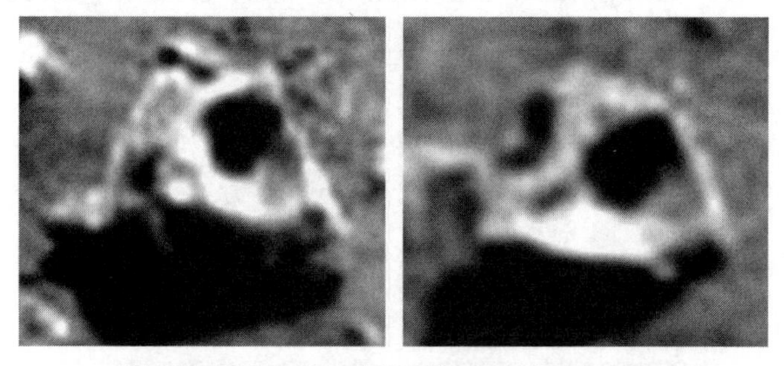

SM31.2 Fossil field? Objects 1 & 2 (detail)

I showed these pictures to a paleontologist who said they did not look like skulls to him and dismissed them. As he is a professional and experienced in identifying such things, I can only respect his learned opinion in this matter. Interestingly, when he enquired about the subject matter of the book, he declared that he did not want to be associated with the project so I haven't mentioned his name.

Most scientists I approached for their opinion on various aspects of this book were polite and helpful and I appreciate this because their professional status and reputation are at stake amongst their colleagues and the scientific community. After all, it could be a disastrous embarrassment were it to become known that a scientist had lent his or her name to a dubious, non-peer-reviewed book about life on Mars.

Anyway, I digress . . . let's get back to this picture.

This whole scene throws up an interesting array of bits and pieces which surely warrant closer examination by experts. I am not qualified to assess whether these objects are fossil fragments or not, as identifying a fossil from a rock is a very specialized field and would almost certainly require an on-site inspection and much closer examination than these pictures can offer. But as amateurs, we can certainly isolate and recognize unusual shapes and therefore pick out something that perhaps does not belong to the field of geology.

Here are some more interesting objects I have highlighted from the main picture, although there are certainly more to be found:

SM31.3 Fossil field? Objects 3-6 (1- 4cm)

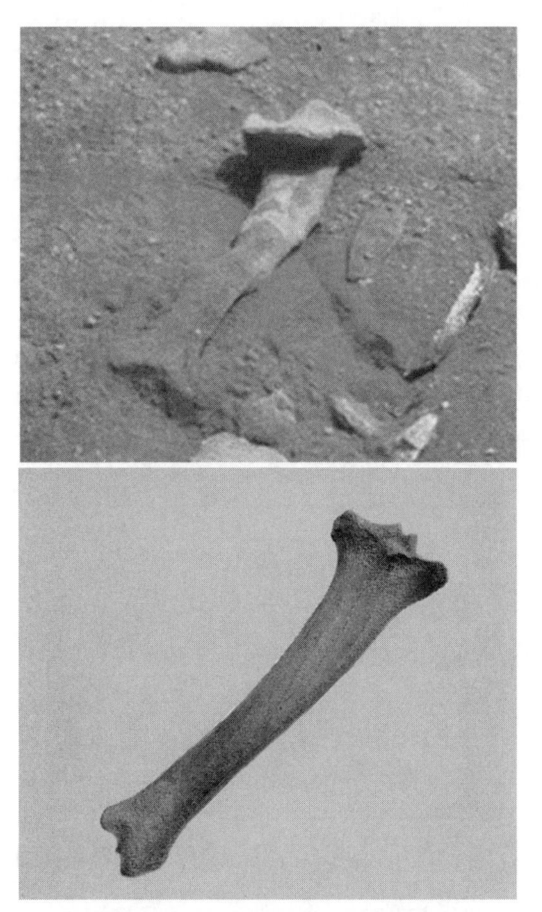

(Top) SM32. Fossilized thigh bone? (10cm) — Curiosity
Region: Gale Crater, Sol 719 Aug 2014
Image credit: NASA/JPL-Caltech/MSSS
(Below) 20. Fossilized thigh bone of a bison (30 cm)—Earth

Potential fossils have also been seen on images from the Curiosity rover, which landed on Mars in 2012. One such example, which resembled a possible thigh bone of an animal (SM32), even received some media attention, but the NASA science team dismissed it as just a rock sculpted by erosion. It was their view that because Mars never had enough oxygen to support large creatures, no such fossils could ever be found.[49] However, to their surprise, *manganese oxide* minerals were discovered in Gale Crater two years later, suggesting that Mars may have had plenty of oxygen to support larger animals.[50] So, dinosaur-bone hunters on Mars just might be in business after all . . . and perhaps these scientists should be more open-minded, for they could have said:

"Well, it certainly *looks* like a bone . . . let's take a closer look". But they didn't, they just drove on. If they had made the effort to analyze it and show it was just a rock, they could have educated the public and raised no suspicion.

SM33. Skull and teeth? (15cm) — Curiosity Rover
Region: Gale Crater, Sol 107 Nov 12; found by Thomas M. S. Jensen
Image credit: NASA/JPL-Caltech/MSSS

Here we have what could be the remains of a skull of some mysterious Martian creature (SM 33). This possibility is raised particularly because we can clearly see a set of what might be teeth protruding from a jawbone. If we are, then this would rather suggest a recently deceased life-form instead of a fossil, because any 'teeth' would not remain exposed or intact in this manner. Of course, it is more likely to be just entangled layers of sedimentary deposits, but let's keep a watchful eye open all the same.

SM34. Animal skeleton? (15cm) — Curiosity Rover
Region: Gale Crater, Sol 109 Nov 2012
Image credit: NASA/JPL-Caltech/MSSS

Here is another potential skeleton or partially fossilized one of an animal, where this time we can see what may be vertebrae laying on the ground, aligned in that familiar parallel fashion (SM34). I estimate this object to be only about 15cm in length, so we could only be talking about a small rodent-sized creature.

SM35. Ammonite fossil? (5-6cm) — Curiosity Rover
Region: Gale Crater, Sol 518 Jan 2014; found by Michael Ivey
Image credits: (Top and lower left) NASA/JPL-Caltech/MSSS, (Lower right) © 2016 Sedgwick Museum of Earth Sciences, U. of Cambridge (reproduced with permission); graphics: SecretMars

I believe this next image, found by Michael Ivey, to be a very strong candidate for a Martian fossil (SM35). As you can see by the picture, the rover wheel of Curiosity came very near to riding over and crushing this very small and potentially very valuable object. True enough when we look at the close-up detail we should be very surprised indeed as to why the Curiosity science team did not attempt to examine it.

Due to its clear distinctive shape, I decided to do an online search for Earth fossils which might be similar to this possible one on Mars, and I was amazed when a picture of an "Uncurled ammonite" finally appeared before my admittedly untrained eyes. But I think you may agree with me that the similarities in shape and characteristics are stunning.

Ammonites were marine mollusks that first appeared on Earth about 240 million years ago until they became extinct at the same time as the dinosaurs. With their distinctive ribbed, spiral-form shells they are the most widely known and commonly found fossils on our planet. As the habitats of these creatures were to be found in the ancient seas of Earth, and as Gale Crater most certainly used to be a water-filled environment—albeit probably a huge lake—we should, I think, begin to accept the possibility that fossil remains will be found there.

Yes, NASA scientists still hold the consensus view that only microbes may ever have lived on Mars and that more complex life probably never evolved at all, but I believe there is more than a fair chance that they will be wrong about that, especially, when we consider the next candidate that I'm going to show you.

In early 2004 a few days after the Opportunity rover had landed on Mars, it was commanded to commence drilling on a nearby outcrop of rock to study its composition. But before it began its work, a few microscopic images were taken of the rock's surface that revealed a potentially astounding discovery—*a possible fossil, similar to an early crinoid from the Earth! (SM36)*

Incredibly, what followed next can only be described as scientific heresy, for three and a half hours later during which time the picture had been viewed by the rover scientists, the abrasion tool was allowed to commence drilling . . . *and it promptly ground the rock face to dust . . . completely destroying what was possibly the most definitive evidence yet that life had existed on Mars.*

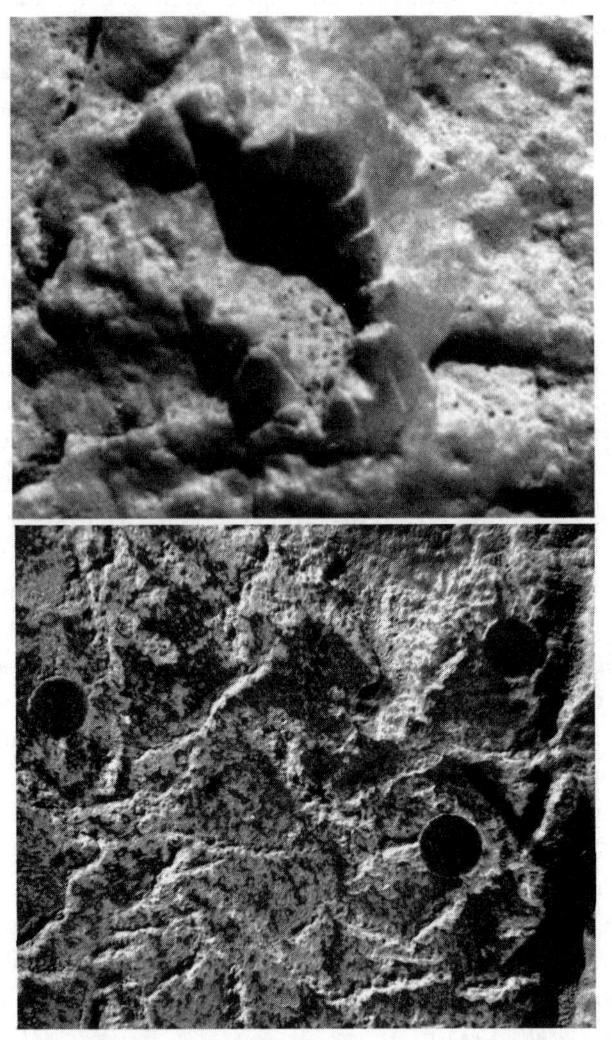

SM36. Possible crinoid fossil destroyed by NASA? (1cm) — Opportunity Rover
Region: Meridiani Planum, Sol 034 Feb 2004
Image credits: NASA/JPL/Cornell/USGS

SM36.1 Earth crinoid fossil comparison

It was researcher Richard C. Hoagland who deserves the credit for both drawing attention to this potential fossil and the extraordinary behavior by the JPL MER Opportunity team. He also tried to persuade the media to take an interest in this story but to little avail.[51] However, more recently in 2014, former NASA scientist Professor Richard B. Hoover took up the story again, when during an interview he explained why he believed that NASA had deliberately destroyed a potential fossil, thus corroborating Hoagland's discovery and claim.

Professor Hoover who retired from NASA in 2011, used to work at the NASA Marshall Space Flight Center on astrobiology and astrophysics. He holds several U.S. patents, has authored numerous papers on astrobiology, extremophiles, solar physics, X-Ray/EUV optics, and meteorites and was NASA inventor of the year in 1992. He has researched microbial extremophiles in the Antarctic, microfossils, and chemical biomarkers in Precambrian rocks and carbonaceous chondrite meteorites. He is more widely known for his claim to have discovered fossilized microorganisms in meteorites. The credentials therefore of this scientist are impeccable and thus what he has to say in this interview is a bombshell:[52]

> **Opportunity rover in 2004 took an image of a fascinating structure on Mars that shows structural features that are consistent with organisms on Earth known as crinoids**. A crinoid is a complicated organism, it's not a bacteria it's an echinoderm [marine animals such as sea urchins and sea lilies found at every ocean depth] like a starfish; we have living echinoderm's on the Earth today.
>
> I spend a lot of time collecting fossils and I've collected a lot of crinoid crowns, this is the head of the crinoid as opposed to the stem, and when I looked at that I immediately recognized it as a crinoid. I showed it to colleagues and friends, I have a friend that has written a book on crinoids and he took one look at it and immediately knew that this was a crinoid.
>
> **Now the fascinating thing is, here you have a possible fossil of a very interesting organism in a rock on Mars and three hours and a half after that photograph was done, that rock was destroyed by the rock abrasion tool**, at least the area where the fossil was located, I should say 'possible' fossil

was located . . . it was ground to dust . . . obviously, a decision was made in some way to utilize the mechanism on the rover to grind away the surface of that object and destroy the structure.

I once asked David McKay [Senior planetary scientist at NASA's Ames Research Center] about this, he explained that he knew all about it and that it was done to look at the inside, looking for carbon or something. Well, the problem is that anyone who does much in the field of paleontology knows that you don't have to find carbon to find a fossil . . . **if a paleontologist finds on Earth a rock containing an interesting fossil, they collect it. You would never have a paleontologist say "Gee, that may represent a new genus of life on Earth. Where's my rock hammer, I want to smash that to bits**". [Emphasis added]

SM36.2 Possible crinoid fossil destroyed by NASA?

So the massive question thrown up here is:

"Why didn't the MER Opportunity team consult with a professional paleontologist to make sure they were not destroying a fossil?"

Was it incompetence? In fact, why was there not a paleontologist employed on the team for this specific contingency? After all, the purpose of all these expensive missions to Mars is to search for life, is it not? And surely they dare not risk destroying any evidence for life that is found by their rovers?

Or perhaps there is something darker going on. Does this event suggest there are people in strategic positions who have instructions to restrict, undermine, or obscure scientific information that might disrupt an already planned agenda?

As far as I know, no one currently employed by NASA on the Curiosity team has gone on record to say that we may find fossils at Gale Crater. This is kind of strange especially when yet another scientist is virtually certain that she has also identified fossils on NASA images sent back to Earth, in this case, microfossils.

21. Rock at Gillespie Lake: ancient home for microfossils? — Curiosity Rover

Region; Gale Crater, Sol 126 Dec 2012

Image credit: NASA/JPL-Caltech/MSSS

Geobiologist Nora Noffke has studied what are called "Microbially-induced sedimentary structures" (MISS) and has become an expert in this field. These structures are currently found in areas of shallow water such as lakes and coastal sites in all locations across the Earth, as well as throughout most of the planet's history. They are comprised of virtual carpets of microbe colonies that can rearrange sediments to form distinctive and recognizable features over a period of time.

Upon studying some pictures from the Curiosity rover at "Gillespie Lake" in Gale Crater (21), she found morphological similarities between Martian sedimentary structures she saw in the images, and microbial structures found at different locations on Earth in Germany, the USA, Africa, and Australia:[53]

> I spent several weeks investigating certain images, centimeter by centimeter, drawing sketches, and comparing them to data from terrestrial structures. And I've worked on these for 20 years, so I knew what to look for.

In her paper published in the journal *Astrobiology*,[54] Noffke described how the structures she found in the Martian rock could also have been formed by salt, water, or wind erosion, but said:

> If the Martian structures aren't of biological origin, then the similarities in morphology, but also in distribution patterns . . . would be an extraordinary coincidence.

She also outlined a plan to confirm the potential biological character of her Martian discovery, one suggestion being to use the SAM instrument on Curiosity to possibly identify important organic or chemical signatures. However, Chris McKay, part of the Curiosity mission's SAM and ChemCam team, argued that because these sedimentary sites on Mars were very ancient, any biological organics would have disappeared a very long time ago.

He also stated that the use of the instrument was now somewhat restricted due to a high contamination level, likely caused during the landing procedure—in essence, that the instrument could not be relied upon to produce accurate results.

What did the rest of the Curiosity team have to say about the discovery of potential microfossils by Nora Noffke in some of their images?

Mission Project Scientist Ashwin Vasavada of NASA/JPL said they did not see anything that could not be explained by natural processes and that in their opinion the rock was just fluvial sandstone.[55] He explained that they did have some members on the team who were keen to keep an eye out for signs of biological processes *(how thoughtful of them...)*, but that they saw no reason in this case to consider anything special about the area or to study it further.

But were these scientists qualified enough in this field to make such an assessment? Did they again make a massive blunder in ignoring yet another sign of ancient life on Mars? What it comes down to is this: did NASA really fly millions of miles to Mars just to ignore a potential goldmine of discovery regarding life on Mars?

Perhaps if they had some paleontologists or geobiologists looking over their shoulder, the existence of ancient life on Mars would already be a fact that we can learn about in our schools, or read up about in our science textbooks, and so begin the expansion of our understanding of life in the universe. But now that we've seen how NASA reacts to potential fossils being discovered by their rovers, we can with reason surmise that they might have an intrinsic problem with the discovery of life on Mars—as we also saw with their reticence in searching for life in liquid water sites, they could also have investigated.

The next category of evidence they refuse to acknowledge as even possible and arbitrarily dismiss any and all evidence presented for it as just 'fantasy and pareidolia'—these are

NASA pictures that potentially show Martian life-forms or artifacts.

Martian Artifact or Biological Life Evidence?

We must presume that all pictures are thoroughly scrutinized. I mean, if NASA is spending billions to send spacecraft to other planets and on expensive cameras to take pictures of where they land, and then have them sent back to Earth for equally expensive processing, archiving and presentation to the public, then we should expect that they must be squeezing every last detail out of them, maximizing what they can learn from these priceless pictures beamed back to Earth from another planet. But are they?

SM37. Debris field? (1.25m) — Spirit Rover
Region; Gusev Crater, Sol 527 June 2005
Image credit: NASA/JPL/Cornell

Here is another picture from the Spirit rover with even more intriguing objects that require a serious explanation (SM37). Grouped together within the space of a meter or so, we find several odd-shaped objects which appear to have little to do with the surrounding rocks. To my eye, they don't 'fit in' with the general character of the rocky terrain and are therefore worthy of our attention—especially the very strange object to the left of the picture (number 1).

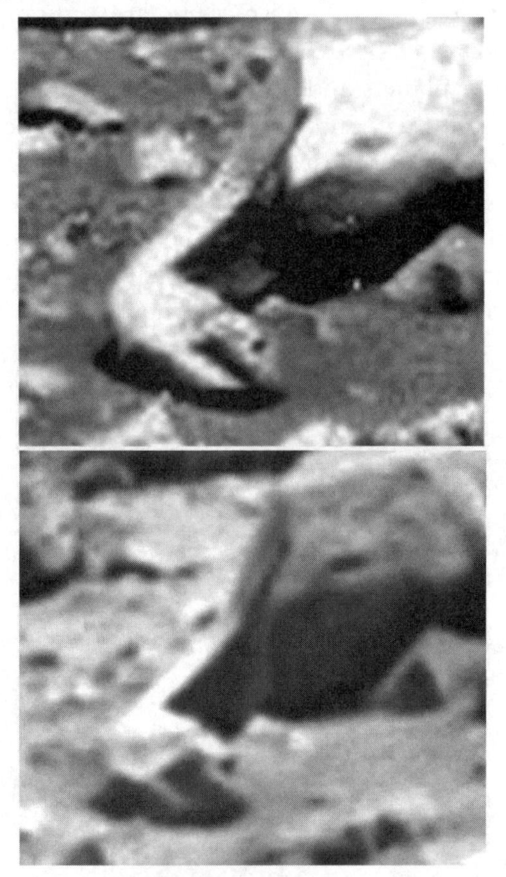

SM37.1 Debris field object 1: the "Eel/Wrench" (20cm)
Found by Rolf Varga

This picture is just fascinating (SM37.1). Are we possibly looking at some sort of creature here? It looks very much like an 'eel', and we could surmise that it may have been expelled from an underground water source onto the surface, where it met its end in the airless and deadly Martian atmosphere.

If not a creature, however, then it might also suggest a twisted, possibly metallic spanner-like object as there is a clearly defined, machined, and angular look at the 'mouth end' of this oddity. Faced with such an amazing object, I went in search of other images to verify its unique shape. This is crucial to confirm whether an object is as mysterious as it seems. Changing the viewing angle as the rover drives by, snapping pictures, can dramatically alter the perspective of an object, enabling a more accurate evaluation of its real shape and character.

In this case, pictures were first taken by the Spirit rover on Sol 527 (June 27th, 2005), which remains the best quality picture, and then a few days later on Sols 540-41 where the viewing angle has changed slightly and the shadows have altered. Later on, a color mosaic was also created by the rover's camera that includes this object (SM37.1, lower picture). These later images confirm the general shape of this eel/spanner-like oddity, but may also suggest an alternative 3D assessment of it too. However, as the better quality image remains the one from Sol 527, we will take that as the more reliable and most accurate one.

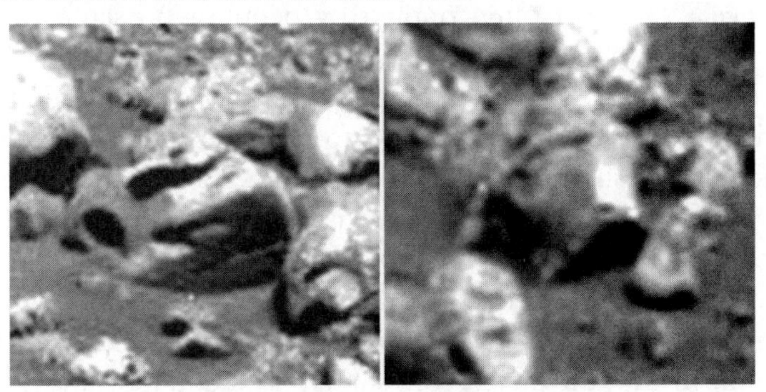

SM37.2 Debris field objects 2-3:
(L) "Lantern" (20cm) & (R) "Casing" (10cm)

The next unusual thing in close proximity to the 'eel' is what appears to be a hollow object with several openings or cavities (SM37.2L). It gives the appearance of certainly being artificial, maybe somewhat twisted from its original shape.

Further off to the right is another curious piece of something which does not fit in with the regular rocks and stones of the surrounding Martian surface. There is the suggestion of a piece of machinery or metallic casing with this one (SM37.2R).

A perusal of this whole picture, taken on Sol 527 of Spirit's exploration of Gusev Crater, reveals an array of bits and pieces that do not resemble regular rocks. To me they suggest a chaotic scattering and dispersal of mechanical fragments from a large technological source—a vehicle or craft—perhaps destroyed by an explosion or a natural disaster such as a torrent of water, for we know that Mars has experienced catastrophic floods in its history.

NASA scientists will certainly acknowledge the possibility of ancient tsunamis having occurred in Mars' past, but are they prepared to consider the existence of *destroyed remnants of technology* on the Martian surface and all the incredible implications such an admission would imply? Carl Sagan once said: "Extraordinary claims require extraordinary evidence"—a retort often quoted by people who scoff at the hypothesis that an advanced civilization may once have existed on Mars. It is just not given serious consideration in scientific circles, and certainly not by NASA; that it is just too outlandish and preposterous an idea. But I would argue, why at *all* should the possible existence of a past civilization on Mars be classed as *extraordinary*?

Most rational thinkers and cosmologists conclude that signs of intelligent extraterrestrial life will eventually be found in the universe. Professor Drake's equation even figured that there could be as many as 20,000 civilizations in our galaxy alone. Now, with NASA's own Kepler Mission having already discovered to date 132 Earth-size planets orbiting near or in the habitable zone considered optimum for life just in our portion of the galaxy, it remains a mystery why NASA and the planetary science community should exclude the planet Mars in our solar system from harboring the possibility that intelligent life may have existed there.

Especially when possible evidence for the existence of that civilization may be staring them in the face through the cameras of their very own Mars rovers…

SM38. Junkyard artifact? (5cm) — Spirit Rover
Region: Gusev Crater, Sol 015 Jan 2004
Image credit: NASA/JPL/Cornell

This is a strange, angular piece of something that would not look out of place in a terrestrial junk-yard (SM38). If not a piece of technology, or something that simply belonged to the rover and has fallen off, or perhaps part of the airbag, it could conceivably be some form of vegetation or perhaps even a deceased life form.

What I find irksome is that if NASA knows what this is, then they must realize that this kind of object is going to raise questions with the public. So I would like to see an information website that focuses intentionally on strange images like this one and which quickly gives the public an explanation for it. Perhaps all the NASA employees on the mission know exactly what it is, but I don't, and I want to.

So let's continue. The following evidence from the old Viking imagery is a bizarre offering which adds another intriguing dimension to this investigation.

SM39. Damaged container oozing liquid, or a life form? (15cm) — Viking Lander 1
Region: Chryse Planitia, May 1977; found by Tim Beech
Source image credit: NASA/JPL; enhancement: Tim Beech

This image could be a trigger for revealing a whole new world on the Martian surface that scientists have not expected (SM39). It was discovered by Tim Beech, with further analyzes done by Joseph Skipper. If the detail in this image can be validated then we may have some highly significant data. My one reservation about it, however, is whether the image enhancement done on the NASA original has indeed revealed a very real 'liquid' or, has created the appearance of it artificially by excessive image processing.

What we have here is a trench dug by one of the scientific arms from the Viking 1 lander after it had scooped up some soil to analyze. About half-way along the trench on the left is what appears to be a 'rock' oozing some kind of light-colored liquid from the side facing the trench. If this is the case, and the image and enhancement correctly shows a liquid, then we can presume that the mechanical arm must have caused this damage by knocking into it.

Further on and to the right of the trench is what might be a fragment that has become detached from the main piece, perhaps dragged along by the arms impetus.

The whole scenario is an intriguing prospect—a 'rock' that has a liquid inside of it—surely highly unusual geology, even for Mars. But if we do indeed have a liquid involved, then almost certainly we are probably dealing with either an artificial container or an unusually shaped form of vegetation or animal life.

Are NASA's rovers trundling over and killing Martian life forms as they pursue their scientific missions for Earthlings? I'm sure NASA astrobiologists would be distraught at the thought, but ignorance will remain just that until eyes are given leave to open a little wider perhaps.

Here's another intriguing example of what may be something organic and living on the surface of Mars, or an artificial object (SM40). There appears to be a curved pipe, a tubular, snake-like extension raised above the surface of the ground that is also possibly connected to a rectangular structure sitting just behind it. There were several images taken of this object which confirm its distinct shape. A rough calculation from the distance to the rover would mean that this strange anomaly could be around a meter in length. It is hard to imagine a more unlikely example of geology. Much rather, it screams at us that it is not.

SM40. Metal pipe or life form? (1m) — Spirit Rover
Region: Gusev Crater, Sol 229 Aug 2004
Image credit: NASA/JPL/Cornell

And what about this strange 'web-footed' little thing (SM41)?

SM41. NASA debris or what? (10cm) — Spirit Rover
Region: Gusev Crater, Sol 36 Feb 2004
Source image credit: NASA/JPL/Cornell; graphic: SecretMars

This odd little geometric shape can be seen very close to the Spirit landing site and the discarded heat shield, so it is therefore almost certainly just a small piece of the craft that became detached or was ejected during the landing procedures, something which I'm sure a NASA engineer will confirm. If not, then what is it?

Before we close this chapter, let's take a quick look at a few mysterious objects widely circulated in the media that have been suggested as evidence of *living, breathing life* on Mars.

Evidence of Life on Mars: Real or Fantasy?

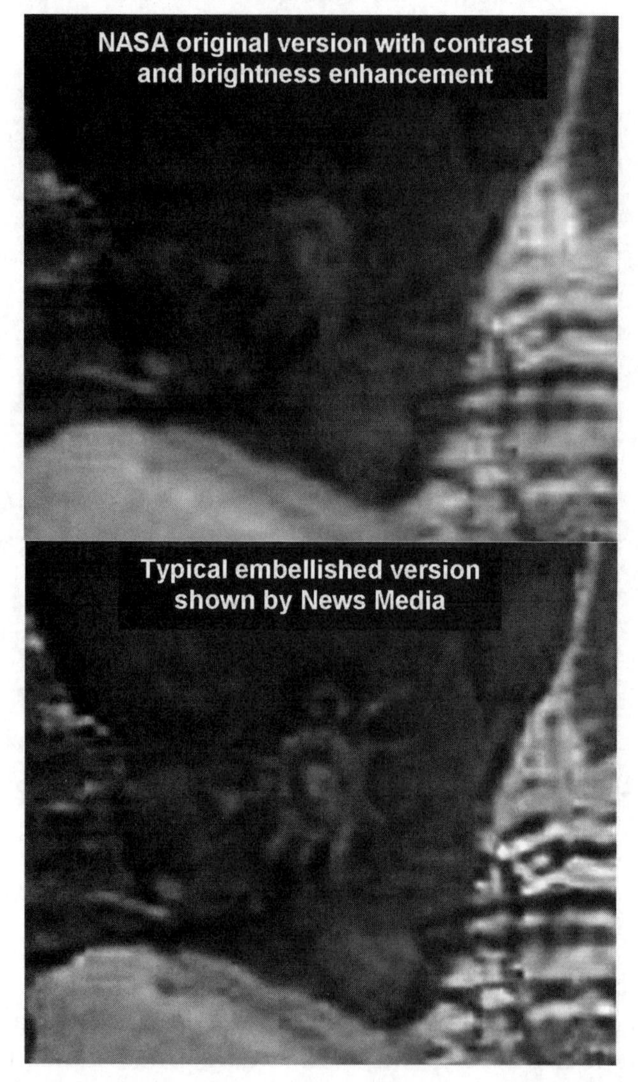

SM42. "The Spider-crab of Mars" (5cm) — Curiosity Rover
Region: Gale Crater, Sol 710 Aug2014; found by Will Farrar
Image credit: NASA/JPL-Caltech/MSSS; graphics: SecretMars

This very strange object was first noticed on an image from the Curiosity rover by researcher Will Farrar. It was later re-processed by others to try and bring out more detail from the picture, after which it then appeared to reveal a bizarre and quite dangerous looking 'spider-crab' creature, complete with several legs and a scorpion-like tail (SM42)!

However, when we have a go at working out the actual size of the thing—and I estimated it to be not much more than 5cm—this fearful image drops considerably on the human danger-level barometer.

But what do we really have here in this NASA picture?

As a potential discovery of an exotic life-form or plant on Mars, the importance of accurate and correct 'image enhancement' really begins to matter, and the question needs to be asked if the detail in this picture was perhaps more so 'embellished' rather than enhanced, possibly exaggerating the limited detail that is there, rather than revealing its true nature.

Because, as we can see from the original NASA image of this thing (SM42.1), there is not much definition or clarity in the picture, and when zooming in close to this object, blocky, linear, JPEG compression artifacts begin to dominate.

Keith Laney a digital imaging and software applications specialist and MOC image processor for the NASA-Ames MOC MER 2003 Landing Sites Project, has critiqued several notorious images that have fallen foul to deliberate exaggeration by certain parties.[56] These were people who were looking to attract media attention and advertising revenue through YouTube by promoting deliberately falsified images.

In short, they knew that by inventing a 'creature' from a NASA image, they would get a rush of public and media interest. Such poor behavior damages the credibility in the public eye of this kind of research, lending fuel to those who want to rubbish the more serious finds.

You can see therefore quite clearly see how the picture of this 'spider-crab' has been artistically fashioned (SM42) from the raw data of the original image (SM42.1). The most disappointing thing is that hidden in the shadow of this rock crevice could well be a mysterious living creature or some kind of vegetation if it's not just an eroded rock feature. Hopefully, the Curiosity rover will revisit this site at a later date and get a closer look at this oddity . . . if it's still there.

SM42.1 "The Spider-crab of Mars" (original NASA image)

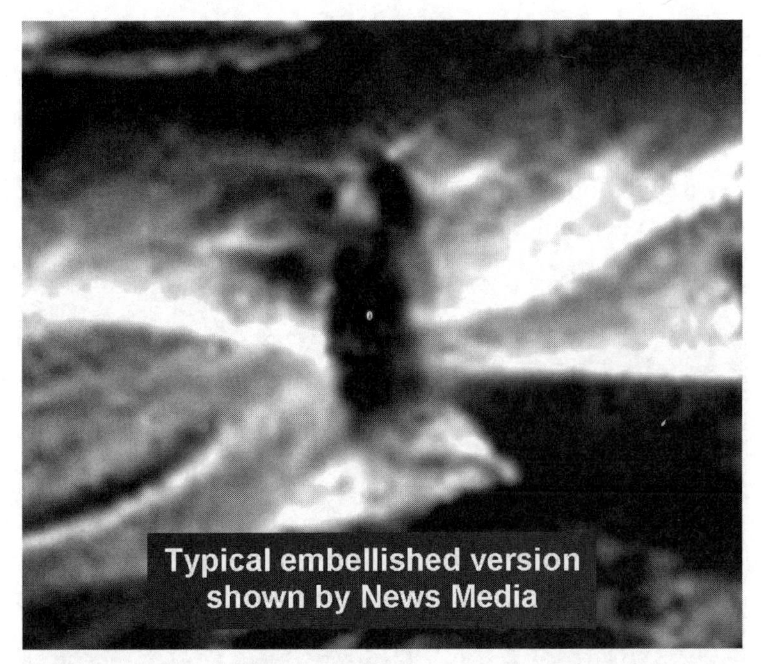

SM43. "The Ghost lady of Mars" (10cm) — Curiosity Rover
Region: Gale Crater, Sol 1001 May 2015
Image credit: NASA/JPL-Caltech/MSSS; graphics: SecretMars

Another picture that received excessive media attention, was a mysterious ghostlike female image (SM43). This appeared widely in fun articles that played on ridiculing the 'paranormal fantasy folk', who of course gullibly believe in such things don't they?

It would be reassuring now and again to see journalists do some genuine research on this subject to provide their readers with some real information rather than pander to creating inane stories. But the puppet strings seem to be pulling in only one direction when it comes to how the media choose to present the subject of life on Mars to the public.

Revealing the truth about this picture is not difficult. As with all the NASA rover pictures, all you need to know is the name of the rover which took the picture (Curiosity) and the Martian day upon which the image was taken (Sol 1001) which you then take to the NASA raw image database.[57] Displayed before you will be all the images taken on that day by the Mastcam, and in time sequence. You are then free to click open and scrutinize every picture until you find what you are looking for.

SM43.1 "The Ghost lady of Mars" (NASA original)

The "Ghost lady" appeared on several images and crucially at different times of day, so that you can track any change in light and shadows, which is key to working out what went on here. As the time sequence of images shows, what we simply have is a rock that changed its appearance as and when the light fell upon it at different times of the day and caused various 'shadow play'. The rest was left up to the human imagination and our natural instinct to identify well-known shapes in objects, especially human features.

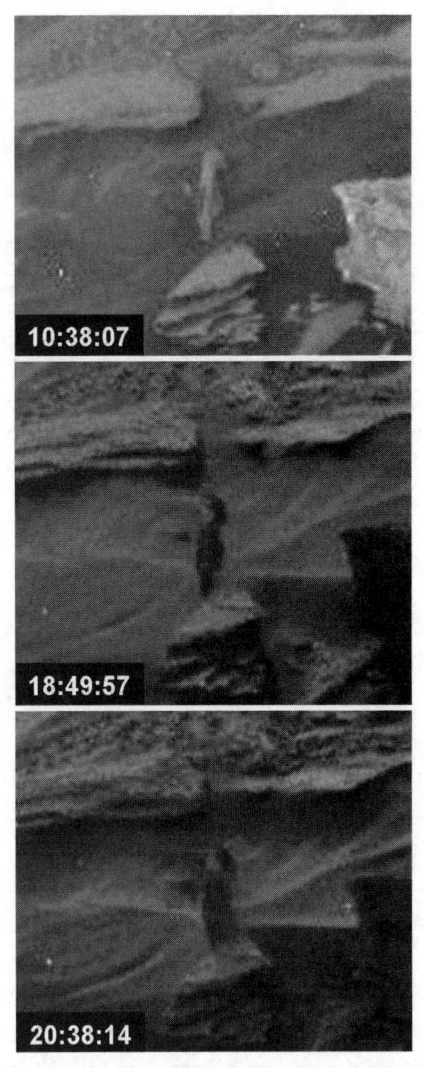

SM43.2 "The Ghost lady of Mars" (time-lapsed)

Here is another object that caused quite a stir when it was first discovered (SM44). Most people saw it as a woman or 'mermaid' sitting daintily perched on the edge of a rock with her arm raised, but . . . once again she must be rather small . . . because you see, the distance from the camera on the Spirit rover is about five meters away, making this object barely *5 cm tall, about 2 inches!*

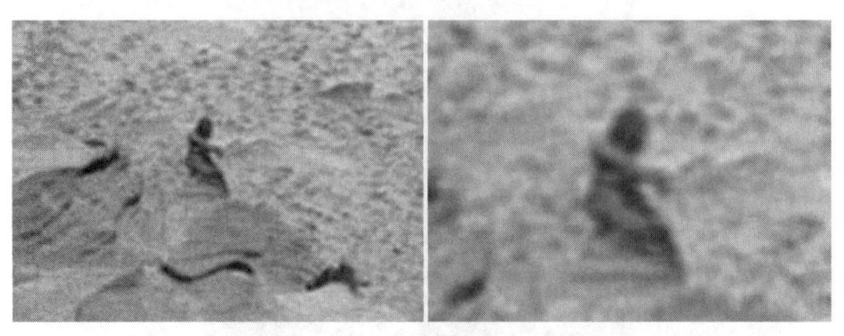

SM44. "The Statuette lady of Mars" (5 cm) — Spirit Rover
Region: Gusev Crater, Sols 1364-67 Nov 07; found by Marcela Bravo
Image credit: NASA/JPL-Caltech/Cornell

Well, some people believe that tiny folk live on Mars, but if that were the case then this object almost certainly isn't one of them. You see for the camera to take this image, the object needed to have stood absolutely still for a full minute or so while the picture was being taken. Reason being that the image is composed of blue, green, and infrared filters and each shot requires a *pause* in between.[58] Furthermore, the Navcam camera onboard Spirit had also taken a shot of this panorama three days previously—and the figure had not moved at all from its position.

This confirms to me that it is not a life form. However, the possibility remains that it could be a small statuette or figurine.

Given what we are about to present to you in the next chapter I would certainly not rule out the possibility that miniature figures or works of art may be found on the surface of Mars. In this case, however, the "Statuette lady of Mars" is probably just a piece of odd-shaped, weathered rock.

Now let's come face to face with the most serious subject of this book: evidence for a civilization that once existed on Mars.

Chapter 4: Signs of Civilization on Mars

Many people, scientists, and academics will be uncomfortable at the thought of a civilization having existed on the Planet Mars because it will upset all of their mainstream preconceived notions about not only the history of that planet but indeed how and when life began and evolved across our entire solar system and even here on the Earth itself. For, if an intelligent civilization once lived on Mars, then there could be profound implications if that civilization had achieved spaceflight, visited our planet Earth, and had bearing on the development of life here.

But if we are discovering pieces of evidence on the Martian surface indicating an artificial origin—and this evidence does not originate from NASA's spacecraft—then we must face the obvious conclusion: an intelligent race on Mars must have been responsible for it being there. Let's see if we can substantiate the hypothesis.

Artificial Objects at Gusev Crater (Spirit)

22. The Spirit Rover landed in Gusev Crater, January 2004 (Artist's impression); image credit: NASA/JPL

We've seen quite a few images taken by the Spirit rover already that appear to have some surprisingly artificial-looking features, but the images I'm about to show you now raise the bar that much higher and invite even more serious consideration that what we are truly looking at is real evidence of mechanical, technological debris on the surface of the planet Mars.

SM45. The "Gear wheel" (10cm) — Spirit Rover
Region: Gusev Crater, Sol 288 Oct 2004; found by Larry Roy
Image credit: NASA/JPL/Cornell

Incredibly, here we have what appears to be a *piece of machinery* half-buried in the sand (SM45). It is about 10 cm in size and looks like a cog mechanism, with familiar, equally spaced teeth extended along the outside of the rim. Some fossils can take a somewhat similar shape, so this too should be considered as a possible explanation, but this seems to me to be a very strong example of a technological past having existed on Mars.

SM46. Metal box or casing? (13cm) — Spirit Rover
Region: Gusev Crater, Sol 1419 Dec 2007; found by Maya
Image credit: NASA/JPL/Cornell

It is just strange that the rover was not directed to examine this object more closely and yet they stop and examine mundane, ordinary-looking rocks they somehow find more 'geologically' interesting.

Accepting that objects with linear, rectangular, and geometric characteristics are more likely to be artificial than geological in origin, let's see what further evidence we can accumulate.

Gusev Crater seems to throw up quite a few anomalous objects. The Spirit rover meandered around another site in this massive 100-mile wide crater to find a very square, hollow, and surely manufactured piece, akin to metallic casing (SM46).

SM47. Manufactured object? (7cm) — Spirit Rover
Region: Gusev Crater, Sol 1402 Dec 2007; found by Kevin Brant
Image credit: NASA/JPL-Caltech

There are also one or two other objects with some geometric regularity, very close to this cube shape. As it appears from the disturbed ground that the rover deliberately drove *over* this area, we have to wonder if perhaps someone at NASA decided to take a closer look at it.

Perhaps we should start to ask what Spirit's true mission was in this crater; I mean, why did NASA land here in the first place? Was there something unusual about this crater? Had some analysis revealed high readings of magnetic activity, a metallic presence, or something mysterious?

And what about this intriguing and amazing little find (SM47)?

We find this one sticking up out of the ground in amongst conventional rocks and totally out of context with the natural geology surrounding it. The angular, geometrically proportionate dimensions and surface sheen are highly suggestive of artificiality.

But, apart from superficially looking like the tail fin of an aircraft, the keel of a boat, or even a metronome, who knows what it is, but it certainly bears the artificial characteristics of an object fashioned by a machine.

And what should we make of this next oddity?

Imagine taking a stroll and you come across a round, curved object lying on the ground (SM48). What would you presume you were looking at? Almost certainly you would stop, reach down and pick it up, thinking it was maybe a coin or something of value.

From this picture, we might also think we were looking at a broken plate or something of that size, but in actual fact, if we compare the rover tread marks in the soil with this object, we can see that it is too small to be a human dinner plate. However, we do need an explanation for the very precise curvature displayed here, as it certainly looks like it could once have been a complete disc.

SM48. Small coin? (2cm) — Spirit Rover
Region: Gusev Crater, Sol 843 May 06; found by Marcus Johannsen
Source image credit: NASA/JPL/Cornell; graphics: SecretMars

Artificial Objects at Gale Crater (Curiosity)

Now let's see what the Curiosity rover has managed to discover since it touched down in Gale Crater, August 2012—although perhaps "discover" is somewhat of a misnomer to describe the massive *lack* of curiosity shown for anything that doesn't look like an obvious rock. This rover, just like the others, seems intent on driving on by, on a blind rush to get somewhere—anywhere it seems—except to stop and investigate the most interesting and unique objects on its path.

When you see the nature of the anomalous objects I am about to present to you, you may well begin to ask "What the hell are they doing over there at NASA? Does anyone even *look* at these pictures?"

23. Mars Science Laboratory Curiosity at Gale Crater (Selfie)
Image credit: NASA/JPL-Caltech

Well, the people who certainly are making an effort to study these images are the independent investigators and researchers at home, who are trying their very best to do the work that NASA archaeologists should be doing. They are not trained in archaeology, but they have trained themselves in spending long hours scouring NASA images for signs of odd-looking objects on the Martian surface. Their eyes have become honed to spotting things that most people just wouldn't see.

For example, there has been sterling work done by researcher Rami Bar Ilan, who has systematically devoted I'm sure well over a thousand hours analyzing Mars images to discover some potentially huge and important finds. And there are so many other intrepid souls too who I'm not aware of that have also given enormous amounts of time to studying these NASA images.

When NASA is finally allowed to admit that the remnants of a Martian civilization probably exist in Gale Crater and elsewhere on Mars, I hope that when the credits start rolling out for this discovery, recognition is given to all those people in the public domain who despite ridicule from NASA and many in the science community, just kept searching, finding and cataloging their discoveries for posterity and humanity.

Some of their discoveries you will see now in the following pages. Of course, some scientists and academics will scoff at their efforts and dismiss every object here as pareidolia or fantasy, but if they are not willing to step up to the mark themselves and take a serious look at these images, then the untrained public will have to do it the best they can.

SM49. The "Wheel hub" (20-25cm) — Curiosity Rover
Region: Gale Crater, Sol 064 Oct 2012
Image credit: NASA/JPL-Caltech/MSSS

I challenge anyone with an open mind and inquisitive nature to dismiss ALL the following objects as 'rocks eroded or carved by nature'. Some may be exactly that, I readily concede that possibility—in fact, *probability*—but let's make no bones about it and be perfectly clear here as to the profound consequences in question regarding this study and investigation: if there is just ONE object amongst this anomalous selection that is artificial or manufactured, then we will need to face the awesome implications of that realization, not shy away from it and pretend it doesn't exist.

We shall begin then with a peculiar formation that I've captioned the "Wheel hub" (SM49). The limitation of the 2D image comes into play here as we would be curious to know if this object is as round as it suggests from this one side that we can see. Perhaps on the other side, it's a completely different shape. It is interesting enough however with its three parallel planes and circular hole, to have surely warranted a visit from the rover to check it out.

SM50. The "Tiny turbine" (5-10 cm) — Curiosity Rover
Region: Gale Crater, Sol 1000 May 2015; found by Rami Bar Ilan
Image credit: NASA/JPL-Caltech/MSSS

This next little object is intriguing (SM50). At most about four inches long, there appear to be three or four radial spokes encircling a turbine shaped structure. The image is not quite sharp enough to confirm these details with any certainty, but very interesting nonetheless.

SM51. Small pyramid (10 cm) — Curiosity Rover
Region: Gale Crater, Sol 978 May 2015; found by Rami Bar Ilan
Image credit: NASA/JPL-Caltech/MSSS

The next image caused a little stir amongst the paranormal media keen to believe that Egyptian-style pyramids were sitting resplendently on planet Mars (SM51). Unfortunately, when we all see a picture of a pyramid, most of us automatically presume that it must be quite big, as we are accustomed to seeing these massive structures on Earth. Furthermore, many articles don't bother to offer any helpful information either for the reader to make a proper judgment, preferring instead to simply excite the 'wow' factor.

The fact is, this pyramid is about four inches tall, although I guess we shouldn't rule out the possibility that what we are seeing is just the tip and the rest of it may be buried underground. Pyramidal forms are however commonly found amongst rocks as they fall, erode, shatter and split into sharp angular shapes across millennia, so although we appear to have quite smooth and symmetrical sides on this little one, we can't see what's behind it and so cannot confirm if it is worthy of further attention at this point, intriguing as it is.

SM52. The "C" (5 cm) — Curiosity Rover
Region: Gale Crater, Sol 817 Nov 2014; found by Rami Bar Ilan
Image credit: NASA/JPL-Caltech/MSSS; graphics: SecretMars

The next object I've decided to show (SM52) is a baffling one, insofar as it's so very close to the Curiosity camera, and the mission scientists must obviously have seen it and discussed it. It's so angular and 'C' shaped and symmetrically proportionate, that I prefer to assume they must have somehow cut this shape deliberately with a laser or something . . .

And now we are about to embark on the surreal. This next one will surely test your credulity, as it still does mine (SM53).

SM53. The "Door bolt" (5cm) — Curiosity Rover
Region: Gale Crater, Sol 440 Nov 2013; found by Michael Ivey
Image credit: NASA/JPL-Caltech/MSSS

For all we know, we are looking at a *small door bolt on Mars*. But, as this is so unlikely, especially as its appearance is so much like a typical door bolt from our planet, I'm persuaded that the explanation must surely be an optical illusion or something. And the reason I say this is because of the way this object is sat in a recess, parallel to the rock—as if the rock had been carved out to accommodate it even. Indeed as if it had been *placed* there!

I did go looking for further images hoping to find a clearer one, perhaps at a different angle with which to compare it, but of the six further images that I did find, the one we are looking at now remains the sharpest and clearest, although the others did at least confirm the objects rectangular shape. Well, I'm not going to rule out anything with this image, but reason tells me this one is perhaps not what it seems. But of course, I may be wrong.

SM54. Wheels and axle? (15cm) — Curiosity Rover
Region: Gale Crater, Sol 729 Aug 14; found by Thomas M. S. Jensen
Image credit: NASA/JPL-Caltech/MSSS; graphics: SecretMars

But now this has surely got to be one of the most amazing finds of all (SM54). Discovered by Thomas Jensen, he came across what can only be described as a *set of wheels connected by an axle!*

If ever we could set nature an impossible task of eroding rock to look like a manufactured object, then surely 'a set of wheels' would be it.

Frustratingly this object is too far away to permit closer examination, in fact, I almost chose not to present it due to the lack of detail. But it is so intriguing and so tantalizing an image that I just had to, especially as there is the strong hint of an inner circle too within the structure of the facing 'wheel', therefore confirming the object's overall resemblance to one.

For sure, it could be just a trick of perspective or the result of image pixels having fun, but there is a suggestion of some other artificial bits and pieces too in this image just behind the wheels, which might strengthen the possibility that what we see in this picture is exactly what it appears to be: a small set of wheels—perhaps similar to those belonging to a child's learning bicycle.

Now that's a thought.

We continue with a truly bizarre triangular-shaped object which culminates into a perfectly circular end (SM55).

It's not very big—only about three inches (7cm)—but why oh why was this thing not investigated by the Curiosity rover? You cannot tell me that they didn't notice this striking shape, and if they know what it is, why then didn't they share this knowledge with the rest of us?

If we are truly trundling over the ancient remains of a civilization that once inhabited Gale Crater, then at some point we might expect to come across building foundations, walls and enclosures, or bits of masonry or statues. Finding such signs of habitation however may be difficult, if there were catastrophic and very destructive forces involved, which may be why we tend to be finding only small bits and pieces that may have belonged to that civilization.

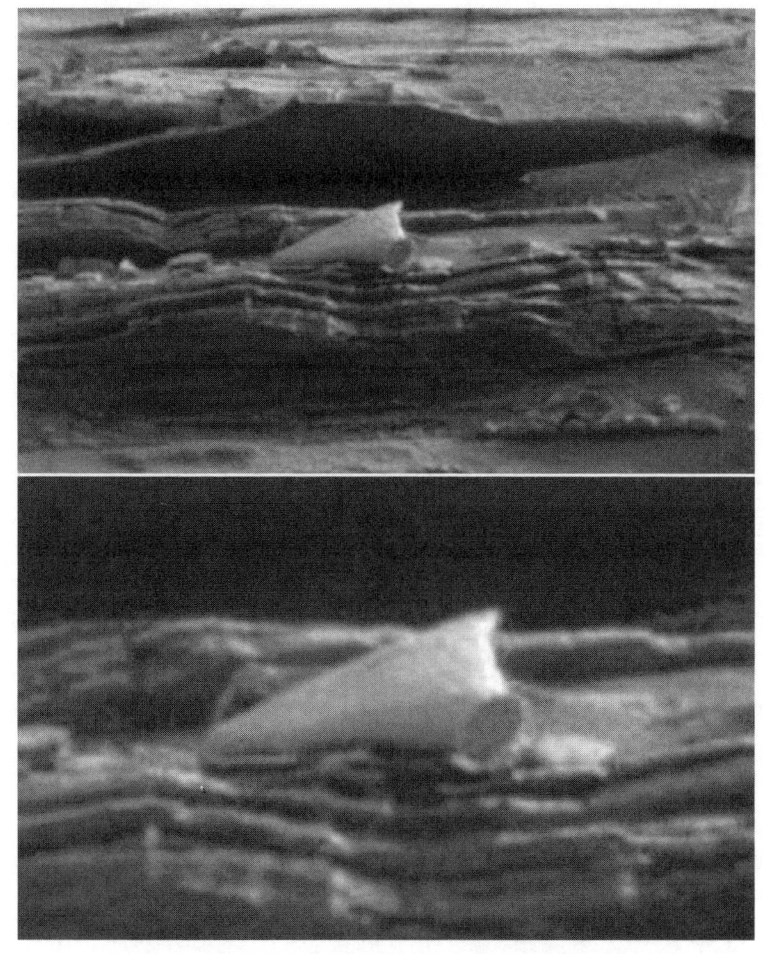

SM55. The "Nozzle" (7cm) — Curiosity Rover
Region: Gale Crater, Sol 821 Nov 2014
Image credit: NASA/JPL-Caltech/MSSS

This next object may well be a possible example of remnant stonework or masonry (SM56) because we can see three possible signs of workmanship here: a symmetrical triangular shape, a parallel plane, and a square end.

SM56. Masonry carving? (25cm) — Curiosity Rover
Region: Gale Crater, Sol 991 May 2015; found by Rami Bar Ilan
Image credit: NASA/JPL-Caltech/MSSS

If utter destruction of a onetime, technologically active Martian settlement occurred, just what kind of things might we find, poking out of the rubble which itself would likely be covered with multiple layers of accumulated dust, maybe also sediments from waters that once inundated the crater and then dried out—a cycle perhaps repeated many times.

Is this the scene that we are exploring today on Mars, left behind by a past era? A chaotic jumble of geology mixed in with remnant artifacts? Often indistinguishable from the rocks and sediment, but occasionally, just now and then, the NASA rover captures an image of a genuine artificial object that has survived the ravages of time, something that was once made and used by real-life Martians a very long time ago?

Something perhaps as innocuous as a mechanical clamp?

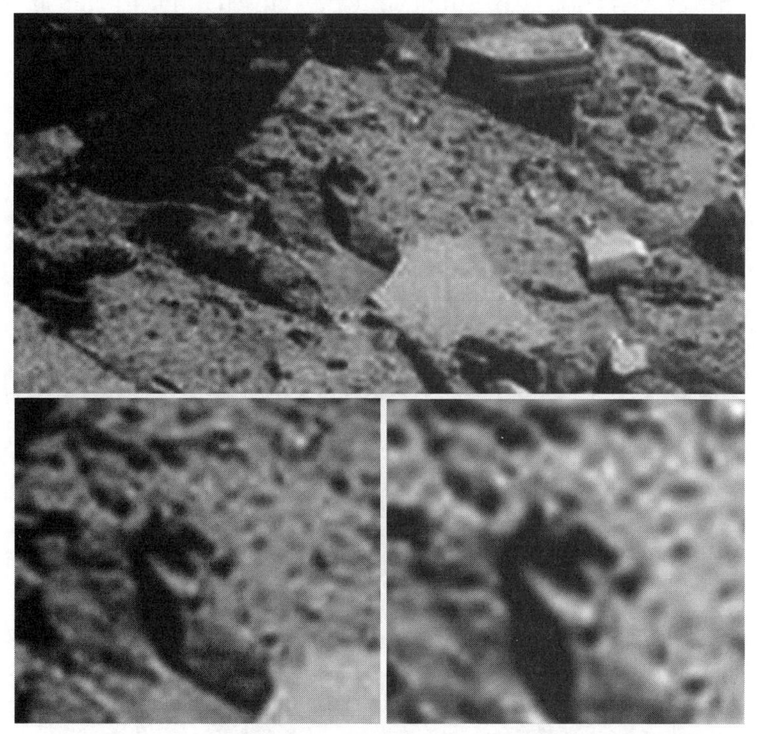

SM57. Beam clamp strut? (5cm) — Curiosity Rover
Region: Gale Crater, Sol 528 Jan 2014
Image credit: NASA/JPL-Caltech/MSSS

I remember when I first saw this object (SM57) in March of 2014, the matter-of-fact feeling it gave me to find something so mundane and functional coming out of the side of a hill on Mars. It simply offered a pragmatic perspective as to what probably happened here, what level of destruction must have occurred to bring this civilization down. And too, a real sense of sadness. Of course, we cannot discount the possibility that we are just looking at an illusion caused by perspective and shadow, but something about this one just strikes a bell.

The next anomalous object (SM58) also has an unmistakable mechanical character about it that most of us will recognize. Yes, both are small, but perhaps telling, in the simple fact of their very existence on the floor of this 96 mile-wide crater on Mars.

SM58. Mechanical fitting? (7cm) — Curiosity Rover
Region: Gale Crater, Sol 109 Nov 2012; found by Michael Ivey
Image credit: NASA/JPL-Caltech/MSSS; graphics: SecretMars

The final two objects I'm going to present from the Curiosity rover's cameras are so explicitly mechanical in appearance and nature that I don't know what else to say about them.

To my mind, the pictures say all that need be said.

The first one (SM59) looks like a machined artifact that has a triangular base with rounded edges. Someone with a mechanical engineering background may recognize more details from this object.

I did search for other images showing it of which there were two, but both were virtually identical so I wasn't able to view this object from a different angle which would have been useful.

SM59. Stanchion base plate? (40cm) — Curiosity Rover
Region: Gale Crater, Sol 595 Apr 2014; found by Rami Bar Ilan
Image credit: NASA/JPL-Caltech/MSSS

The last object is much easier to describe (SM60). It's so astounding it just beggars belief that someone did not command the rover to immediately turn around and check it out.

It does not take much imagination to see a crank handle mechanism here. The close-up reveals far too many intricate, geometric details to come to any other conclusion.

What it is exactly, or what its function may have been, is not important at this stage. That will be for on-site archaeologists to investigate, examine, and determine at some future date. What matters now is that this is either a device or, an extraordinary piece of eroded rock or sediment. I leave you, the reader, to be the judge.

SM60. Crank handle device? (15cm) — Curiosity Rover
Region: Gale Crater, Sol 1051 July 2015
Found by Rami Bar Ilan/Gerald Turner
Image source: NASA/JPL-Caltech/MSSS; graphics: SecretMars

Based on the foregoing pictures, I think we can confidently say the Curiosity mission has one stubborn parameter ingrained firmly into its programming: to simply ignore and drive past *everything* that might be an artificial object.

Not to focus any attention on it, except in such cases where it can easily be dismissed as an illusion or weather erosion, such as a blurry picture of a rock that looks like a ghostly woman.

Martian/Human Tech Debris or Just Geology?

It should be considered a possibility that these scattered objects, if not somehow belonging to the current operational spacecraft on the surface of Mars, could also be the remains of previous failed American and Russian missions that have crashed or broken up and disintegrated in the atmosphere. This would necessitate that some pieces have impacted the surface intact, in which case perhaps designers and engineers of those spacecraft might be able to identify the bits we have drawn attention to here and solve a few of our mysteries.

We must also remain aware that geology certainly does create geometric, artificial-looking shapes and therefore we cannot discount the possibility that some, or indeed most of these objects, may just be random natural oddities. Perhaps geologists can present some natural examples of what we have shown here to help us out in this regard. There is not much else to consider regards explaining these mysterious objects apart from camera imaging irregularities and optical illusions.

The scattered variety of potential artificial objects we have seen indicates that a destructive force may have been responsible for their random and chaotic dispersal; a natural or technological event or disaster perhaps being the cause (or, if Martians were anything like the people of Earth, the result of war?).

To my mind, the most reasonable and rational explanation available is that most of these objects originated from Mars. The rovers drove by and their cameras imaged the remains of a Martian civilization that are still visible, half-buried in the soil *(what lays beneath the surface intrigues the imagi-*

nation). However, there is one observation that bothers me a lot about these objects and that is the *location* in which they have been found. And it is this question we must ask:

What is the likelihood that so many examples of artificial-looking objects have been discovered *along the paths taken by the NASA rovers?*

These vehicles have traversed and taken pictures of only a very tiny area on the Red Planet, and yet we have come across so many examples of what appear to be geometrically shaped manufactured objects. If this is a typical snapshot of the Martian surface then the amount of evidence for a technologically advanced past, strewn across the entire globe, must be colossal. Either that or, if such evidence is rare, then NASA must have surely been extraordinarily 'lucky' to have chosen these landing sites for their spacecraft and then to have stumbled across such evidence.

An alternative argument would be that the very fact such objects are found almost everywhere the rovers drive, shows that they can *only* be unusual rocks because the coincidence would be just too much. A logical deduction but one which assumes there was never an ancient civilization on Mars, to begin with, and hence, no artifacts to be found.

Another alternative, of course, is that someone at NASA has *targeted these landing areas specifically* because they expected to find something extraordinary there. That, following analysis, these areas had been found to display a high order of distinctiveness worthy of investigation such as having high levels of metallic or magnetic properties. But if that is the case then they are certainly not discussing with us what they are truly finding and examining with the Mars rovers.

The ramifications of such a disturbing possibility are of course huge because this would suggest there is a *two-tier investigation* on Mars:

1. A ***mundane*** mission: to focus on geology, climate, ancient water, and microbial life—the results of which

are made available to the mainstream science community and for public consumption.

2. A ***stealth*** mission: to discover the true reality and history of life on Mars and its technological past—the results of which are being withheld from scientists and the public.

This is not a leap of speculation but an assessment that so happens to fit the facts and observations we are now beginning to make of the Mars Exploration Program. Something is not right. Some extraordinary evidence is being scientifically ignored and we have to ask "Why"? If vigorous and healthy curiosity is said to be at the forefront of scientific space exploration, then why is it so blatantly absent here?

We have to ask, how can our scientists become all excited and drooling at the mouth at the prospect of drilling into another rock, and yet, when an artificial cog mechanism or an obvious device with a crank handle sits there on the Martian surface, all curiosity and zeal diminish to a deafening silence?

Or, is it simply that when an operator sits there open-mouthed at their monitor screen, staring at some amazing artificial object poking out of the Martian sand, someone just taps them on the shoulder . . .

SM61. Earth/Mars comparison: artifacts

Let's continue to build the casefile. Here's an interesting image from the Opportunity rover that just might suggest we are looking at an artificial construction (SM62).

SM62. Martian cairn? (1m) — Opportunity Rover
Region: Meridiani Planum, Sol 2467 Jan 2011
Image credit: NASA/JPL/Cornell

It looks like the ruins of one of those South American pyramids but in a more worn down condition. The fact that just to the left of the larger object is what seems to be the remains of another similar and far more eroded version lends support to this possibility.

However, given the distance from the rover camera, it can only be miniscule in comparison, perhaps a meter or so high. Also, the protrusions sticking up from the top of the object look very similar to some volcanic rocks commonly seen on Mars, so it is probably much more likely to be one of those than an artificial construction, with natural weathering processes likely to have caused the pyramidal form.

However, the shape is quite linear and there are similarities between it and those ancient rock cairns built by various tribes in the British Isles, so I would not entirely rule out the possibility that this may indeed be a constructed artifact, perhaps a stone marker of some kind.

SM63. Building foundation? (6-7m) — Curiosity Rover
Region: Gale Crater, Sol 528 Jan 2014
Image credit: NASA/JPL-Caltech/MSSS

Here is an interesting terrace-like feature seen at "Dingo Gap" (SM63) that may be the foundation for a building long destroyed. Perhaps significantly, it is situated in the same area where the small clamp-like object was found, in addition to several other curious objects that are not featured in this book. Could this location on Mars, therefore, mark the site of an ancient settlement? If so, we can only wonder what caused its utter destruction.

Well, we've now seen some small-scale evidence that Mars may have been inhabited by an intelligent species capable of creating and making things: a small coin, a square box, some machine parts, and maybe some small monuments and a stone terrace, too, at least. We've also realized that finding objects with geometrically proportioned and linear characteristics provide us with a possible defining signature of something that has been made by an intelligent hand.

It is by no means conclusive evidence, as geology will certainly throw up geometrically shaped rocks from time to time and we must also keep in mind the optical illusions that images will trick us with now and then; the play of shadow, light and viewing perspective, etc.

We must also be aware of the psychological phenomenon of pareidolia that we've alluded to now and again. This is when the human brain decides to attach significance to random shapes, thinking that "If it *looks like* a tea plate, then it *must be* a tea plate". Overall, this is why I prefer to place more significance on several unusual objects which are found in proximity to each other as this strengthens the likelihood of them having a common source of origin rather than a random, natural one.

OK, the more evidence we find with the telling signature of geometry the more serious our body of evidence will grow, so pursuing this rationale—that straight, geometrically proportioned formations might suggest an artificial origin—we will now look at some objects that follow a *linear course and direction* rather than characteristics of the object itself.

Now, nature is quite capable of manifesting exceedingly straight lines through a variety of geological processes, so we mustn't get too excited whenever we see them. What will determine 'artificiality' for our purposes, will be the context in which the 'straightness' appears; a strong degree of uniqueness that perhaps cannot easily be explained by geology.

So, what are the chances of finding three, similar-sized and shaped spherical objects in sequence, 240 meters apart, on the planet Mars (SM64)?

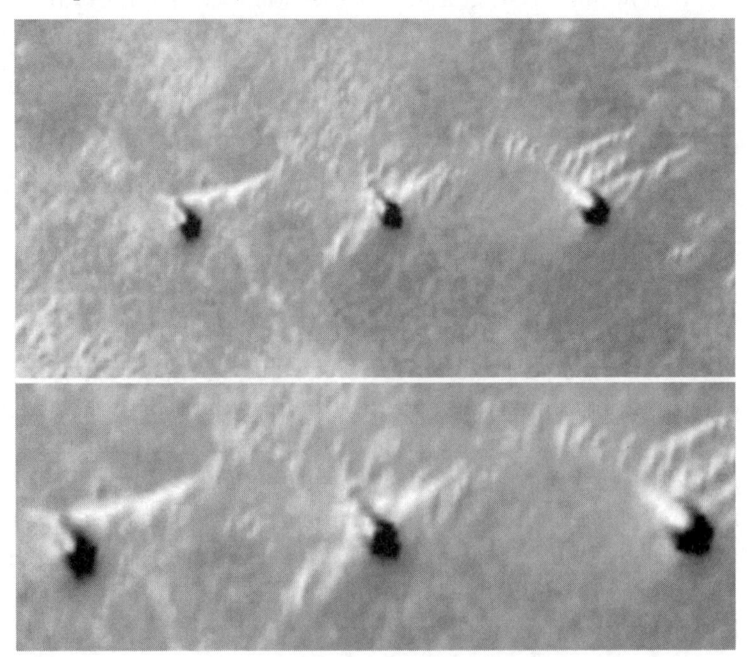

SM64. Natural or artificial mounds? (each object 20-25m) — MOC/MGS Orbiter

Region: Terra Meridiani, Apr 1999; image credit: NASA/JPL/MSSS

Although I have no idea what these objects are and even though they appear suspiciously artificial, I'm pretty sure that nature can arrange something geological to achieve this display, so we won't allow ourselves to get too excited. However, what is the chance of finding *thirteen or more* objects equally spaced apart on the same planet? I would suggest that might be asking a bit too much of nature. . .

SM65. Trail left by the rover, or something else? (25m) — Opportunity Rover
Region: Meridiani Planum, Sol 1070 Jan 2007
Image credit: NASA/JPL-Caltech

Close to the horizon line of this picture we see a string of very similar objects spaced equidistantly apart. To the right of the line the first few objects are extremely alike in shape and size, while further off to the left, they become more elongated and varied in their appearance.

Now there should be a perfectly logical explanation for this and because of the straight line formation the first thing that springs to mind, of course, is the Opportunity rover itself: it could simply be a trail left as it passed that way. Also perhaps, it could be the markings left by a small meteorite that may have skimmed the surface, digging up these protrusions before it came to a halt.

Another rational explanation is perhaps an impact trail left by the airbags when the spacecraft landed—but this is ruled out when we discover the picture was taken at Victoria Crater, on Opportunity's mission day, Sol 1070. This means the rover had long left its landing site over a thousand Martian days ago.

Overall I think the most likely explanation is that the trail came from the rover itself. The camera did take three photos of this area just beyond the crater, but no mention was made of these markings in their mission summary for the day.[59]

If the sensible alternatives fail, trying to work out what's going on here is going to be a puzzle. We would have to consider more extreme possibilities such as some technology devices that are fixed to the ground, or a moving procession of vehicles or even living creatures! But reason tells me we should favor a trail created by the rover itself. The fact is that NASA must know what it is, as it's so blatant and impossible to ignore, so a mundane explanation is a virtual certainty. I did query this image at the NASA website, but no one has replied with an answer yet.

However, if, in the extreme unlikelihood that NASA doesn't already know what these objects are, we need to ask them why they didn't send the rover to find out.

We began this trail of evidence by first looking at some small things that were found close to the rovers and then to somewhat larger objects viewed at a distance from them. We will now move to a scale that will require us to switch from the onboard cameras of the Mars rovers to the cameras of NASA's orbiting spacecraft.

Large-Scale Artificial Martian Constructions

Now that we have discovered what appear to be small pieces of machinery and the remains of constructed artifacts laying in the Martian soil, what are the chances of finding *larger* pieces elsewhere on the planet?

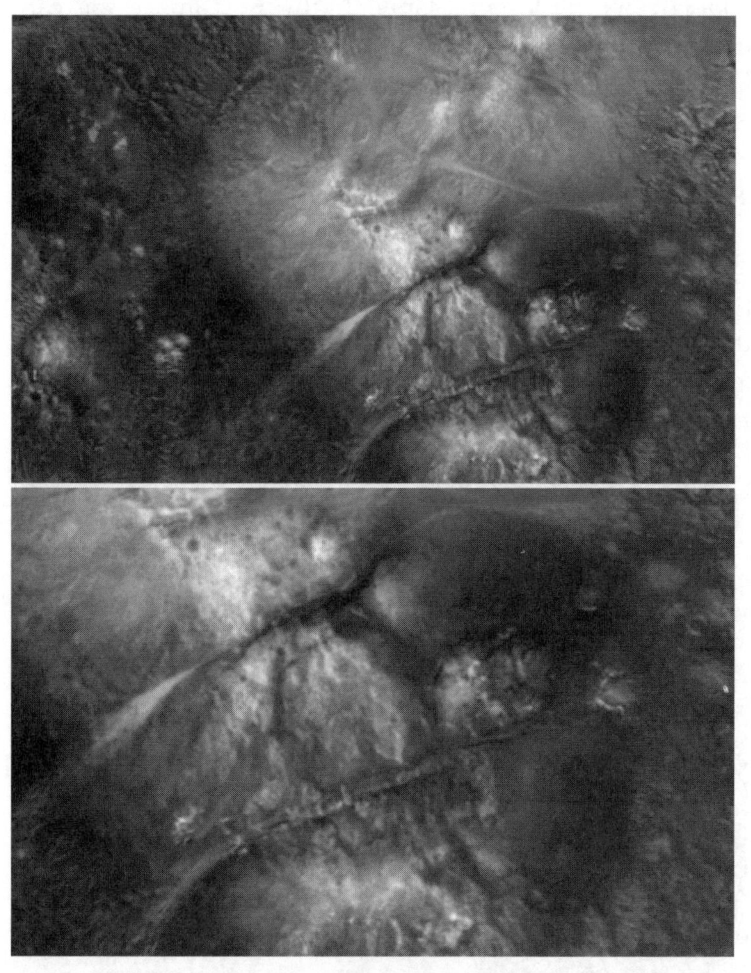

SM66. Large-scale pipe constructions? (4.5km) — MOC/MGS Orbiter
Region: Aram Chaos, Jan 2000; found by J. P. Skipper
Image credit: NASA/JPL/MSSS

If there was once a civilization on Mars that suffered planetary destruction, could there still be some intact remnants of their technology? What an incentive for NASA and its contractors to get to Mars: the discovery of ancient Martian technology!

Discovered by Joseph Skipper,[60] this next image was captured by the MOC (Mars Orbiting Camera) from Mars Global Surveyor and is an extraordinary picture (SM66). It is a baffling image to comprehend and one teeters towards seeing either technology or some complex, bizarre Martian geology.

What strikes one here are the two 'pipeline' structures that appear to be pouring liquid out through a nozzle. The upper pipeline seems to be caught in the very act of spraying the liquid (water?) if indeed this image is actually showing us an elevated 'spray' and is not merely a ground-level illusion.

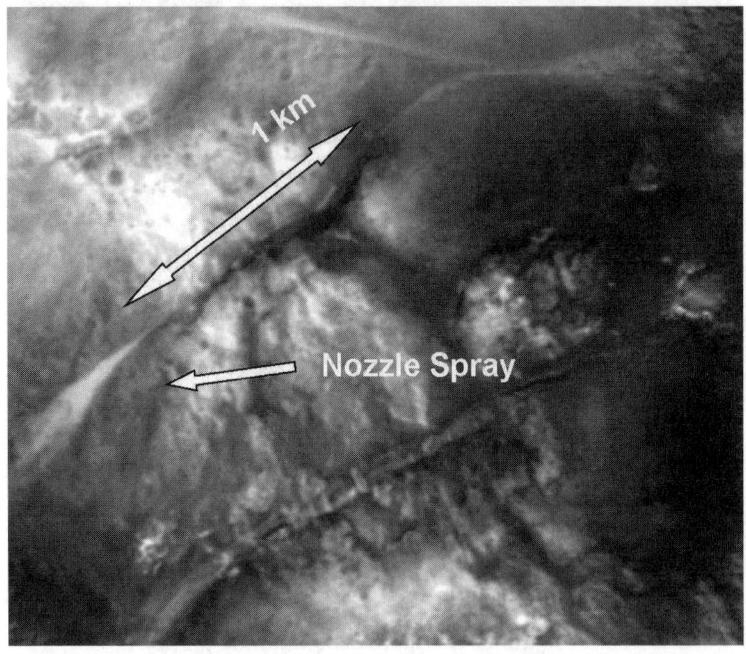

SM66.1 Detail of "Pipes"

If we observe a little closer, Skipper suggests that the nozzles appear to be connected to possible 'sacks or domes'. The lower pipeline's sack looking somewhat 'deflated' compared to the upper one. We can also see a bit further up from these 'two pipelines' the suggestion of perhaps a third one that has 'dried up' or no longer functions. These features are much less certain, though.

Just how big are these structures? The MOC image strip has a width of 2.91 km so we can calculate that the nozzles are about 1km or more in length and the nozzle end about 15 meters in diameter. That would be a powerful, thunderous torrent pouring from the pipe; a frenetic freezing and boiling froth disgorging into the Martian air.

This, of course, could all be nonsense if what we are looking at is just another strange-looking, geological oddity. To me, there does seem to be an organic, natural feel to the objects in this image. What then could explain these pipe-like formations?

I wonder if somehow, over long periods of time, a source of subterranean water, perhaps an upwelling, is constantly making its way to the surface and, due to the atmospheric conditions on Mars, has formed a chain of geological steps that have resulted in these 'pipelines'. The water will boil away or freeze upon being exposed to the Martian air, so therefore these pipes could be natural structures formed over a prolonged period by constantly replenished water ice, or maybe mineral deposits.

I tend to lean towards a geological explanation, but as there appear to be several structured elements in the vicinity of these 'pipes' I have some doubt. If technological, one may presume it to be long-abandoned, but perhaps still functioning, its purpose having been to just draw upon a natural cycle in operation here, like tapping a well. And even though not in use the water still comes to the surface and follows its course in this bizarre fashion.

It is all wild conjecture, of course, but this is some intriguing evidence to consider and I think it is, therefore, a worthwhile exercise to speculate a bit. We should, however, keep in mind that there is sufficient ambiguity in this image to throw significant doubt as to what we are seeing here; that what appear to be 'pipelines' and 'domes' may well be ground-level illusions caused by perspective and image resolution. It would be wise therefore to keep tabs on this one until more evidence comes to light as to the true nature of this scene.

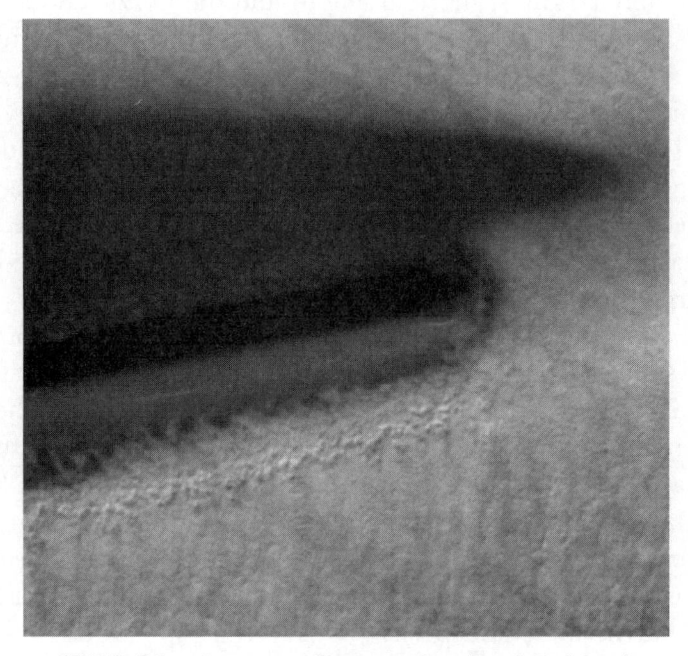

SM67. Martian tunneling mechanism? (1.9 km) — MOC/MGS Orbiter
Region: South Polar, Feb 2000; found by J. P. Skipper
Image credit: NASA/JPL/MSSS

Now, if that is unusual, this next piece of evidence for a technological presence on Mars is positively surreal (SM67). I really hesitated to include this one because I suspected it was just an illusion caused by perspective or something. But have a look at it yourself and see what you think.

Again brought to my attention by Skipper,[61] this formation is described in the official MSSS image caption as a

"Sample trough with dark floor in South Polar terrain". Now, I've turned this image upside down and sideways to somehow see the straight line as the scientists do—but I'm not convinced.

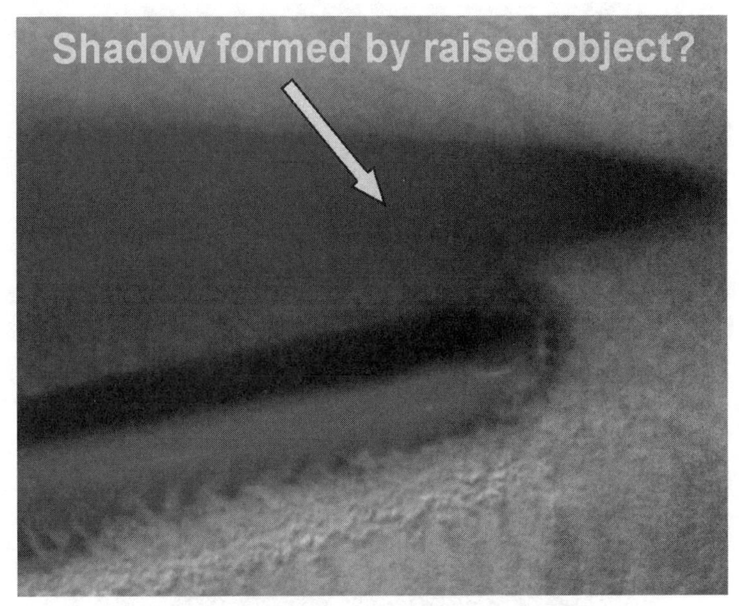

SM67.1 Detail of object and "Shadow"

The guys at MSSS are saying that the straight dark feature is a 'trough'—a depression in the ground. But from the *shadow* cast beyond it (or is it a shadow?) this might suggest that it is actually an object extending *upwards* and *above* the surface. It can also give the impression that it is some kind of pipe or tube burrowing its way *into* the ground!

I would cautiously dismiss this as a possible artificial object, were it not for the stunningly straight appearance of this feature. And if we are indeed seeing something that extends above the surface, and not as the geologists say, a trough that is carved into it, then we could conceivably be looking at some kind of mechanical device here, perhaps engaged in the excavation of the Martian surface.

This tentative possibility, of course, throws up the incredible question: are there intelligent beings on Mars today? Well, if what we are seeing here is a correct observation, then…

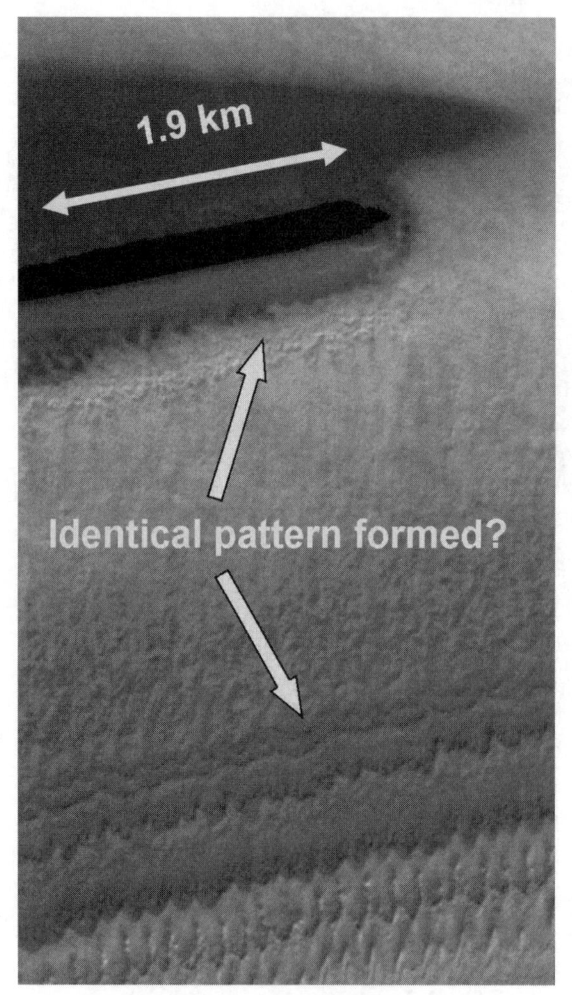

SM67.2 Detail of pattern and size

Further down from the dark, straight feature we see parallel markings on the ground. This pattern may suggest that the 'device' has already been there and done its work.

Could it be laying some kind of subsurface tube or tunnel system? The absolute straightness, the light-reflective sheen of the surface, and the beveled front end are indicative of an artificial, manufactured device. The shape of the object is not dissimilar to a modern high-speed train, but I wonder.

If you've ever read Frank Herbert's *Dune* trilogy you might begin to ponder whether that author perhaps had a premonition of Mars. Such intuition is not uncommon in creative human beings. The story focused on a dry, desert planet that had huge hidden reserves of water beneath the surface, and an indigenous life form—giant subterranean worms that now and then came up to the surface to roam across the planet.

Well, stranger things will be discovered on Mars in the future, of that much I'm certain.

The idea that an intelligent civilization used to exist on this planet is at least a conceivable one in most rational minds, however, the possibility that there are living beings inhabiting the world *now* in the present day is a subject that few serious space scientists with a reputation to protect will engage in or discuss on a public platform. And yet, these genuine NASA images we've seen may well be pointing to such a reality on the Red Planet.

Scattered pieces of technology may suggest a far distant past when Mars used to have a living civilization going about its business, but the existence of these objects may also point to a much more recent past for that civilization. It depends on how severe natural erosive forces on Mars can be.

Artifacts may not be able to remain identifiable for very long on the surface, meaning that any relics discovered would have to be fairly recent. Then again, perhaps shifting Martian winds periodically uncover and then bury this evidence, which means they could be very old indeed. It's hard to say.

What Happened to the Martians?

24. Was all hope for life evolving on Mars, extinguished by huge cosmic impacts?

That Mars has experienced a catastrophic past is not questioned by any scientist. The enormous asteroid impact craters, the cavernous rips and scars in the Martian surface, the lava plains, the scouring caused by massive torrents of water bear testimony to the almighty and devastating destruction of what was certainly a potential life-bearing world—and maybe one of great beauty—utterly shattered by massive and terrible cosmic events.

Although some believe that Mars experienced a gradual and slow death over many millions of years it seems more likely that the asteroid impacts it received were the probable cause of the planet's extinction; a devastating blow from which it could not recover. Its magnetic field damaged, its atmosphere dissipated, and its ability to sustain life perhaps almost entirely extinguished.

If a civilization ever existed on Mars we can only imagine the nightmare it must have endured as all signs of it were exterminated from the planet's surface: incinerating walls of flame; gigantic tidal waves as her oceans heaved and swayed from the huge surface impact; massively destructive earthquakes and rivers of lava consuming all that moved, with gigantic missiles of rock raining down on everything that lived.

After the horrors of the planet's destruction passed, what would be left? The surface would be in darkness as volcanic clouds would hide the sun and freeze the surface for years after. Deadly radiation would bathe the planet as the protective atmosphere and ozone layer would have burnt away or been lost to space.

Where Did the Martians Go?

A Martian people may have been aware of the coming catastrophe and had some time to prepare. They could have gone underground, perhaps hiding or fortifying themselves in deep caves. They could also have had the means to leave the planet and journey to another world in the solar system, perhaps to the Earth, where there are many ancient legends of beings arriving from space and bringing teachings and influence to different races.

If they remained on the planet some of them may be still there, living deep underground. And if we discover images showing technological constructions still in operation then this may be evidence of their continuing existence on the planet. They could also all have died a very long time ago.

However, if they had left the planet and came to the Earth then their ancestors may be walking amongst us today. Maybe we are the Martians.

And I think it is this possibility which might explain why a lid is being kept on the truth about Mars. A complete rewriting of Earth's prehistory is a very uncomfortable idea, and the impact on human religions—especially those that consider the Earth to still be the center of the universe, and humanity the Creator's only project involving life and intelligence in the universe—would be thrown into confusion.

Where Did the Martians Live?

Whether all the Martians died, stayed on their planet and went underground, or managed to leave their doomed world and escape to another, one irrefutable sign of their former existence should at least remain . . . some scattered ruins of their civilization. The question is, what could we possibly find after such enormous destruction?

We have found already some significant evidence of small-scale debris, and perhaps larger constructions too, that suggest there was a technologically advanced race living on Mars, but what about evidence of where they lived, such as roads and cities? Surely we should be able to see some signs of their habitation.

The reality is that most of these would surely have been obliterated by the cataclysm that destroyed Mars, and over time, natural weathering would have covered anything left over with multiple layers of dust and sand. The ferocious Martian winds too would have severely eroded any walls and buildings that had remained intact. Perhaps what still survives of any Martian civilization simply depends on just how long ago this all happened.

I recall seeing a fascinating documentary on British television a few years ago which predicted what might remain of human civilization were we to suddenly disappear or die out; [62] in effect, how many years would it take before nature would erode all remnants of human technology and buildings from the face of the Earth and render our civilized achievements invisible to any visitor to our planet.

The sobering figure arrived at was just 10,000 years.

In geological time, that is a miniscule number of years. Interestingly, the film showed how the only buildings that may just be recognizable as constructed artifacts after 10,000 years of Earth's weather and natural erosion would be the Great Pyramid of Giza in Egypt, some parts of the Great Wall of China, and . . . the granite-carved sculpture of Mount Rushmore.

25. The Pyramids in Egypt: after a global catastrophe would this be all that remains of human civilization?

If we add the forces of time and erosion to planetary destruction on Mars, can we expect to find anything at all left of a civilization? If all traces of the human race would disappear after 10,000 years, what will be found of a Martian civilization that may have perished 100,000 years ago, or even up to a million years or longer? What state would any remaining ruins be in after such a long time of being exposed to the Martian elements?

The reality is that the chance of finding intact remains of an ancient Martian civilization's buildings and structures on the surface of the planet must be remote. We might just hope that remains may still be found in isolated areas that perhaps by chance experienced less destruction than the rest of the planet and where the elements may have helped to partially expose buried cities and habitats. Also that some isolated fragments may still exist, such as the small-scale evidence we found in the form of possible technology parts and debris seen by the cameras of the Mars rovers.

We must hope that, just as winds will cover the ancient remains of a building with soil and sand, they can also expose some of that building by eroding the soil away. We will now study some images that may be showing us evidence of where the Martian peoples used to live; the final remains of a civilization that may have been destroyed by a terrible cosmic cataclysm.

Satellite Archaeology on Mars

Imagine visitors approaching the planet Earth from space, arriving in their spacecraft to our blue, white-clouded world. How would they know that an intelligent civilization lived here? The first indication of intelligent life on Earth to these space-faring travelers would probably become apparent when the *geometric regularity of our constructions* came into view on their monitor screens.

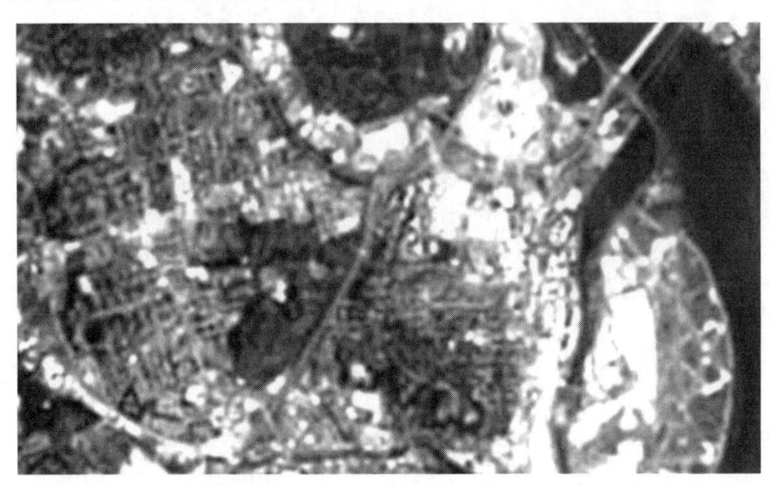

26. Extensive geometrical patterns reveal the presence of intelligently created artificial complexes—satellite image of Washington D.C. in the United States
Image credit: NASA/Landsat

As a signpost for what we might come to recognize as ancient sites of civilization on Mars, we will be looking for these same revealing geometric signatures that tell us when we have correctly identified an artificial construct here on the Earth. Also, we will be comparing partially buried habitats that have been identified by archaeologists here on our own planet with potential Martian ruins to see if there are correlations. Interestingly—and NASA, take note—many archaeological sites on Earth have solely been discovered and identified through studies of *aerial photography and satellite imagery*.

Archaeologists on Earth routinely use aircraft to survey, identify, and map hidden sites—a technique they have relied on for decades. But since satellite imagery became widely available in recent years, and especially with the launch of *Google Earth* in 2005, these scientists have been able to scan for ancient ruins over large areas of land covering our entire planet.

We don't yet have aircraft that can fly over Mars to search for ruins, but we certainly do have satellite imagery—as we have seen in this book. NASA has taken over a hundred thousand images of the surface of Mars, many in high-resolution detail, and I would be extremely surprised if someone somewhere in some official capacity, has not studied these images for signs of civilization.

The reality is that it would be unbelievable foolishness not to have done so. For that reason I am certain that someone somewhere has had the sense and intention to do it. I have no proof that NASA or some other office is conducting an archaeological survey of Mars, but we can apply reason to a situation to uncover a strong likelihood, and my sense of reason tells me that this research is being done. The problem is of course that they are not talking about it or discussing this aspect of research openly in public.

Martian archaeology is being kept a secret.

This is a great shame and will cause long-lasting distrust when eventually this information is revealed. Preferably the authorities will come clean soon and allow the public and the science community to embrace this search too, but if they won't then we will just have to discover and pursue the truth ourselves, as we have done with the evidence already revealed so far in this book.

I hope that this public effort will eventually force the authorities to acquiesce and recognize that the tax-payers are several steps ahead of the game and won't be taken for fools or accept the scarce drip-feed of information that they are told is the limit of truth and knowledge available to them.

Therefore, in this matter of possible archaeological evidence existing on Mars, I call on professionals the world over who are experienced in aerial searches for ancient buried sites on the Earth to turn their expert attention towards Mars and space archaeology to see what they can discover.

Geometry in Martian Archaeology

To reiterate: in order to identify possible sites of Martian ruins, we will only be able to rely on *one* set of invaluable criteria. An indisputable signature that will almost always reveal the presence of intelligent thinking and design: geometry.

If we can discover simple, geometric features on the surface of Mars, then we will have a fair chance of revealing sites where Martian civilization used to exist. Features that show us straight lines in relation to right angles and rectangular, repeating patterns that will likely indicate homes and habitats, streets and buildings buried beneath the Martian sands.

These can often be identified from aerial studies of images on Earth because underlying patterns just below the surface can often only be recognized as such from the air. As we said, thousands of new archaeological sites have been discovered since Google Earth allowed armchair archaeologists to study satellite images of our own planet, and new technologies involving lasers and infrared technology have revealed hidden sites invisible to the human eye. There is nothing like regular, repeating geometry to signify the presence of intelligent design and layout.

It could be argued however that an alien civilization may build in a *non*-geometrical fashion, perhaps preferring to mimic nature or geology instead of rectilinear structures as we do here on the Earth. In which case we will never be able to identify their buildings whatever we do. In this regard, therefore, we can only proceed with our investigation on the assumption that intelligent beings in our universe will be aware of the same mathematical constants that we ourselves have discovered, and have adopted a similar approach to building and architectural design.

So, for this study, we will look for straight lines and right-angled formations; geometrically constant signatures that stand apart from natural, geological landforms. For the most part, what we find will probably fall somewhere in between the two and will leave us with the question that we have had to face throughout most of the evidence presented in this book: is it natural or artificial?

As ever, you and I will have to make the judgment.

The exciting thing though is that if we do discover clear evidence of rectilinear structures on Mars we will surely have irrefutable, archaeologically acceptable evidence that an ancient civilization once existed there . . . or will we?

Geologists will point out that there are many examples of angular and geometric formations on Mars, and here on the Earth, that are created by natural geological forces, and most us have seen them too: we could cite "Giant's Causeway" in Ireland, where hexagonal vertical blocks sit side by side, formed by lava; look also at the shapes of crystals. What this should remind us of is that geometry is also an intrinsic part of nature—that there is something in the matrix of matter that determines these forms and shapes to manifest in our universe.

However, evidence of geometry, as manifested by *a civilization*, is on a different plane of order, perhaps even inferior—and with respect to the human race, often certainly so. What we build is not always beautiful or organically woven into nature; all too often, rather an unsightly splurge, grafted onto the land with careless insensitivity.

Ironically, this difference may be the key to identifying the geometry of artificial structures as opposed to those of natural origin—especially with regards to the crucial factor of *location*. For example, a civilization will build forts and strongholds on hills and mesas so it can watch out for enemies that might invade and destroy it. Or it centers on coastal sites where it can fish or build ports for ships to facilitate trade with other folk from far-off lands.

Such considerations do not exist within nature's geological confines but they certainly do within the minds of thinking, conscious beings concerned with self-preservation, and the survival and continuation of their species.

How Advanced was Martian Civilization?

If we have indeed correctly discovered and identified small remnants of technology on the Martian surface in the images we have so far presented as evidence for a Martian civilization then we can expect at least a civilization that had reached Earth's industrial revolution stage. If large-scale technological devices have correctly been identified, then probably beyond that and approaching our twentieth-century level.

As with our planet, where we discover buildings and ruins that stretch back over four thousand years of human civilization—from the ancient Mesopotamian, Egyptian, Indian and Chinese empires, through to the Greek, Roman, and Mayan—Martian civilization may also too have remains from older times in its history, and what we may discover in NASA images of their civilization, could therefore conceivably come from any period stretching back across many thousands of years of Martian history, not necessarily remnants from their most advanced society.

Interestingly too, we find many ruins on the Earth that have withstood the ravages of time simply because they are out of the way and have remained relatively unmolested. This due to the fact they were built on high points for safety and security—hilltops, mountain mesas, and plateaus—and where also perhaps, weather conditions and erosion factors were less severe.

Modern Earth civilization now builds its towns on level plains near rivers and coastal sites rather than in the hills. If planetary devastation caused these low-lying centers of civilization to be completely obliterated, perhaps the higher-placed habitats—and these more ancient ruins, no longer inhabited—may indeed also be the most common of remaining ruins to be found on Mars.

For example, a planetary catastrophe may completely destroy and bury London and New York through giant tsunamis and earthquakes. Many years later an alien ship visiting the Earth and searching for signs of human civilization may discover only the remains of places such as Masada in Israel or Machu Picchu, Peru, perched up high on mountains, long deserted by living souls, yet the sole remnant of visible human civilization to visitors from space.

27. Machu Picchu — Peru, South America

The one fact too that we must keep in mind and which is of enormous consequence is that we need only find *one* remnant site of *any* Martian civilization—just one piece of conclusive evidence that leaves us in no doubt at all that Mars used to have intelligent people living on it. Because once we do that, then we will be forced to ask the question "How did it get there?" and then the whole edifice of NASA's current scientific investigation of Mars will be shown to be a shameful fraud and their guise that "Only real science like geology and astrobiology should be done on Mars" reduced to an embarrassed choke and splutter.

When I first began this search for possible evidence of Martian civilization I reasoned that it would make sense to narrow the search parameters and keep a particular eye open for possible Martian hill habitats, forts, and places that may once have been coastal towns. The latter being possible because scientists have suggested the theoretical location of ancient seas, oceans, and lakes on Mars and the shorelines that once lapped these.

However, as I looked through the available evidence discovered so far by independent researchers that claimed to be possible sites of ancient Martian towns and buildings, I realized that it was a major task indeed to find credible evidence of ancient habitats at all, regardless of where they were located. But our internet mavericks certainly found some as you will soon see for yourself. And a few are truly remarkable. So let's begin.

The following images were brought to my attention via leading researcher and exponent for ancient civilization evidence on Mars, Richard C. Hoagland. His discoveries of possible locations for Martian ruins constitute the best evidence available, as far as I've seen.[63]

Here is an elevated feature that could be a candidate for a hidden ruin (SM68). We see two very clear right-angled formations that mirror their opposite corners. Such clear geometry would suggest we may be looking at walls and the severely eroded remains of a square-shaped building or enclosure of some kind.

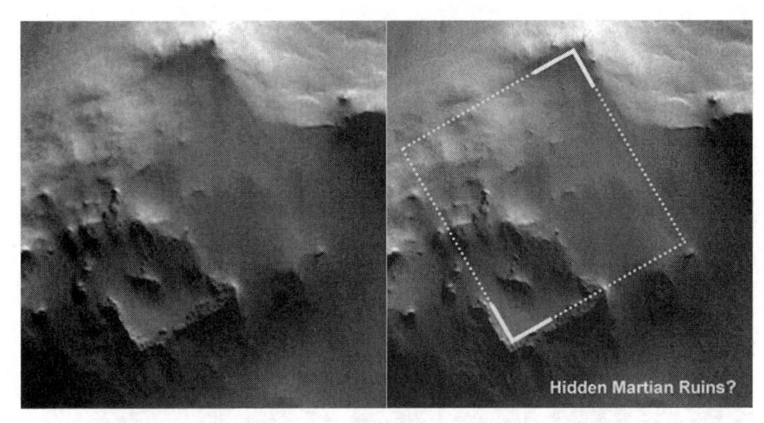

SM68. Ancient Martian ruins? (1.7 km) — MOC/MGS
Region: Arabia Terra, Nov 2001; found by R. C. Hoagland
Image credit: NASA/JPL/MSSS; graphics: SecretMars

As a comparison, let's have a look at an archaeological site at Sarvistan in Iran to see how ruins will sometimes appear from the air (28):

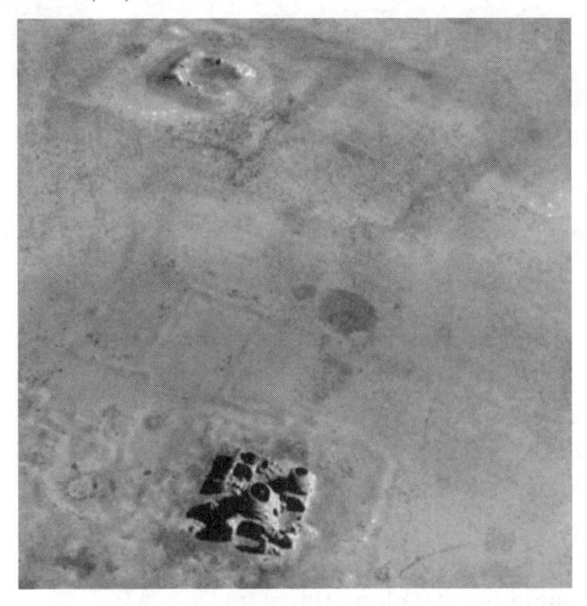

28. Earth ruins, near Sarvistan, Iran — 1936
Image credit: Oriental Institute, University of Chicago.
(Reproduced with permission)

The structure seen at the foot of this photo is what remains of the Sasanian Palace. However, what interests us most about this scene is towards the top of the image, where we can see square markings on the ground indicating the former, or buried presence of ancient walls and enclosures that still leave a trace on the surface and which can clearly be seen from the air.

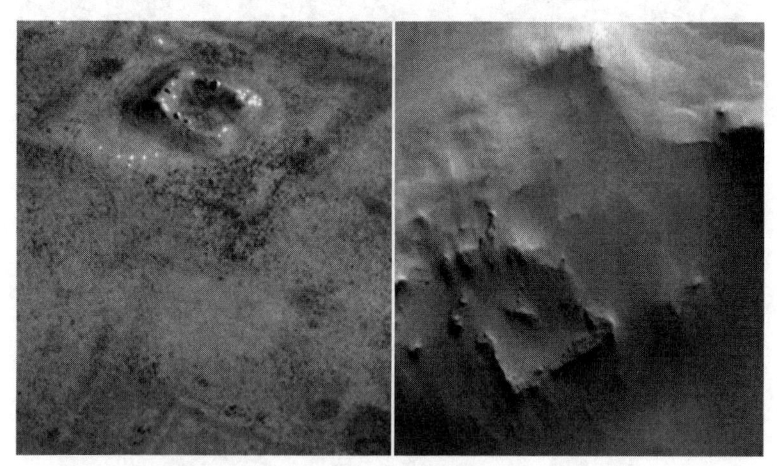

SM68.1 (L) Earth ruins in Iran (R) Buried ruins on Mars?

Now let's take the upper portion of this photo that shows these markings of former walls and enclosures and compare them side by side with the possible ruins at Arabia Terra on Mars (SM68.1). The photo from Iran was taken from an altitude of only some 425 meters and is on a smaller scale than the Martian structure, but the fundamental purpose of this comparison remains valid in that it shows that if we find rectangular and linear formations here on the ground on Earth, then we will presume that there is an underlying archaeological explanation for them—that a building used to exist there; one that was originally constructed by intelligent, thinking beings a long time ago.

Now, are we prepared to extend this rationale that applies to discoveries found on Earth, all the way to the planet

Mars? Be prepared for what is coming. This evidence I believe is truly extraordinary.

We are now back on Mars and are going to take a close look at a section of an image taken by the HiRISE camera from the Mars Odyssey orbiter, back in 2006. What we are seeing now is an example of what would normally be presumed to be a crater off to the left of the picture, along with some seemingly unremarkable geological terrain off to the right (SM69). The clarity is not as sharp as we would have liked in these images, but there is more than enough detail here for us to make some definitive observations and conclusions.

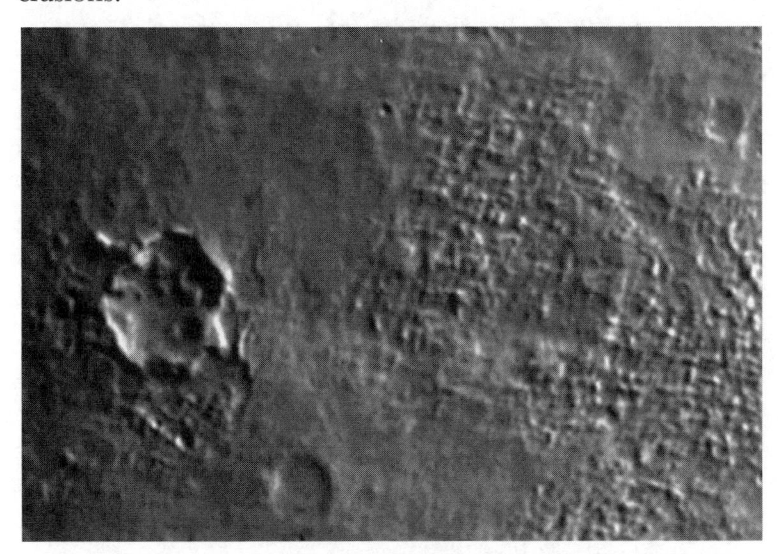

SM69. Partially buried remains of Martian village? (1 km)
HiRISE/MRO Orbiter
Region: Mid-latitude southern highlands, Mar 2006
Found by R C Hoagland; image credit: NASA/JPL-Caltech/UA

Let's take a closer look at this picture then and zoom-in on the bumpy, nodular landscape on the right-hand side (SM69.1), where we do not have to scrutinize too carefully before we begin to see a pattern of consistent, rectilinear structure here—a repeating grid that gives the appearance of *regular-spaced and grouped buildings; what may well be the remains of a Martian village, half-buried in the soil . . .*

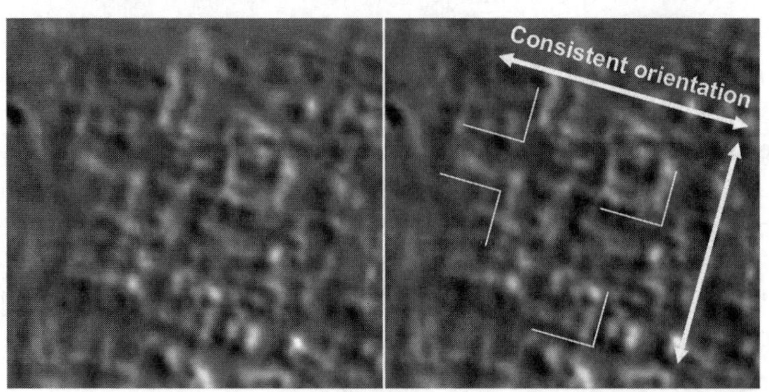

SM69.1 Close-up view of nodular terrain (200m/656 ft across)

We can quite clearly see repeating patterns of right-angled structures very akin to buildings, walls, and enclosures as would be laid out in typical grid fashion in a human settlement. Perhaps your imagination can, as did mine, see the foundations of walls that used to be rooms belonging to houses, along with space for paths and walkways; a flight of fancy maybe, or not?

But to proceed further with this hypothesis we need to first ask an important question and discover what the ground-scale is here on these remote Martian sands. Is it of sufficient size to fit in whole buildings and streets? I mean, if this location was only 10 meters across we would be talking about the land of Lilliput!

Well, I calculated as carefully as I could and found that this exact area in question (SM69.2) measures about 200 meters across. Then, after some scrutiny, I began to realize (with quite an eerie feeling actually) that several houses of human size with several rooms each, could quite comfortably fit here.

Now that could be significant, presuming of course that any Martians who once lived here were similar in size to us.

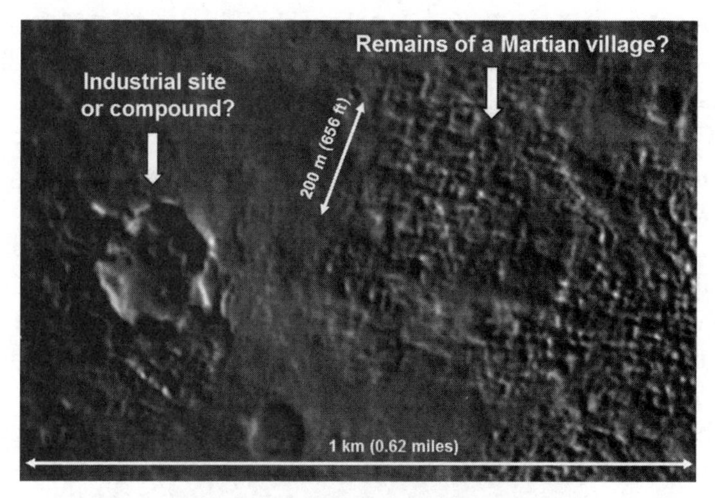

SM69.2 Ground scale of "Village" and "Compound"

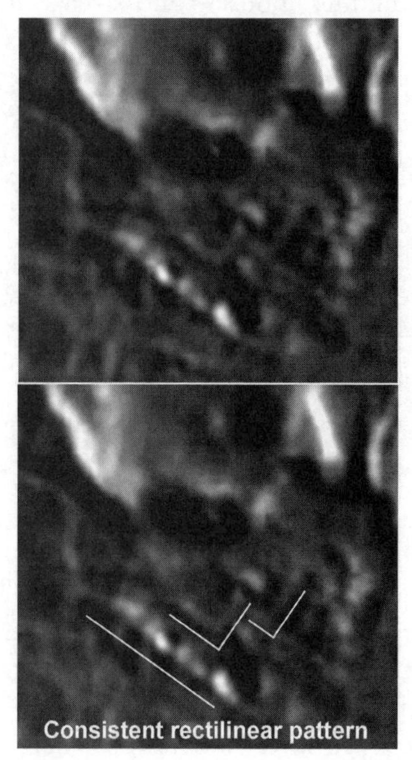

SM69.3 Rectilinear detail south of "Compound"

Well, as we continue to explore this site we can observe this linear pattern continue and extend around to the oddly shaped 'crater' off to the left of the main image (SM69.3).

Can geology mimic this geometric pattern? It certainly can to some degree, but to the extent, we are seeing here, I have my doubts.

To give us an idea of what half-buried buildings and city walls appear like on the Earth and compare the probability that we are seeing the same kind of underlying structures here on Mars, let's look at a photo taken of the ancient archaeological site of Ashur in Iraq (SM69.4).

Here in the photo on the left (L), we see a section of the ruins at Ashur, with once again, the very obvious rectilinear layout that is very hard to ignore. We are seeing here partially buried buildings, houses, and streets that have been excavated by archaeologists at the site.

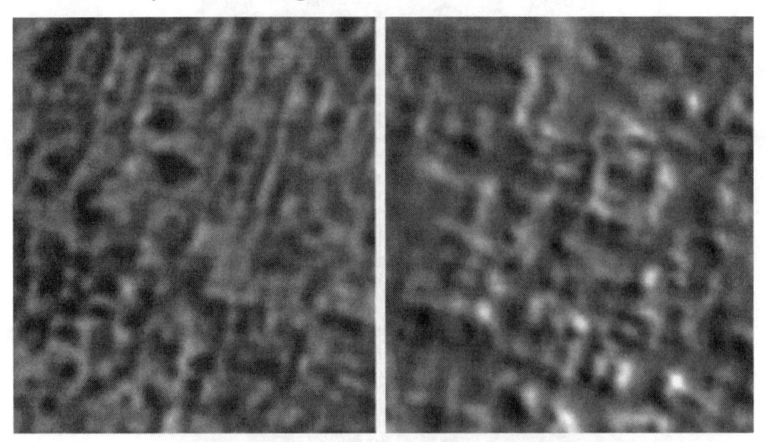

SM69.4 (L) Town ruins in Ashur, Iraq
(R) Town ruins on Mars?
Image credit (L): ©2013 DigitalGlobe, Google Earth

When we compare this real-life archaeological location on Earth with the image from Mars (R), the geometric similarity and regularity are just striking.

Now let's move back to the 'crater', where we are in for even more of a surprise.

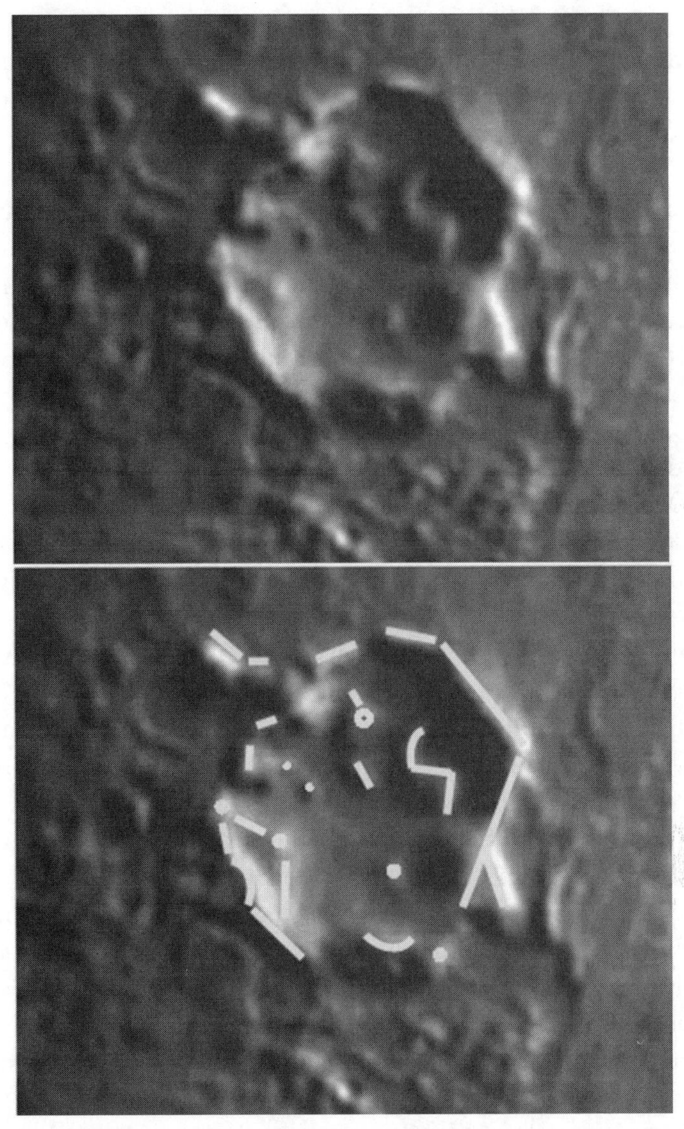

SM69.5 Detail: "Compound/industrial site" (200m/656 ft)

As we get closer to this lop-sided structure, we can see that it looks absolutely nothing like a crater at all and much more like an artificial construction and enclosure (SM69.5). This is particularly apparent when we notice the diverse array of objects within its walls, where we can see curved, straight, and right-angled formations all sharing the same locality. The whole complex measures about 195-200 meters across.

This whole scene displays a very strong signature to me that we are probably looking at constructs and not geology, as there appear to be just far too many elements indicative of deliberate design, concentrated within a relatively small area. I would also suggest that this enclosure and the objects within it appear to display an architectural relationship with each other too; a recognizable layout and intrinsic harmony. If validated this would of course vastly increase the odds that we are indeed looking at a work of artificial origin.

Overall, I would consider this whole site encompassing the 'crater' enclosure and the 'town/village' area to be a very strong candidate for evidence of a civilization having lived on Mars, of which these partial ruins remain. Furthermore, I am certain this site is known about by those at NASA or elsewhere who are looking specifically for this kind of evidence. The sober reality, however, and something we must always keep in mind, is that they have deliberately chosen not to tell us about it and to keep this information secret.

Will NASA land their next spacecraft here so they can take a look? Will they set aside their pedantic search for ancient water and microbes and instead go and give us some *real* evidence with which we can consider our place in the universe? Well, this will depend on whether NASA will extend its scientific parameters to include the science of archaeology in their investigation of Mars. Publicly, the subject of possible evidence for an ancient civilization on Mars is just not on the table. They won't discuss it. Well, we had better help them get talking about it.

Let's move on.

At this point, it is worth considering what kind of structures or buildings, mounds of sand, or bumps on the surface might be hiding beneath. In Egypt, lost cities were buried for a thousand years by the desert sands before archaeologists became curious and dug them out. We dare not blind ourselves with presumption if we desire to make real discoveries, whether here on Earth or other worlds—and in this particular adventure, we must also dare ourselves to imagine what wonders may lay hidden just below the Martian surface.

In this next case, however, we're not looking at something obscured by sand, but rather a possible sign of civilization which seems to be quite revealing as to its structure and definition.

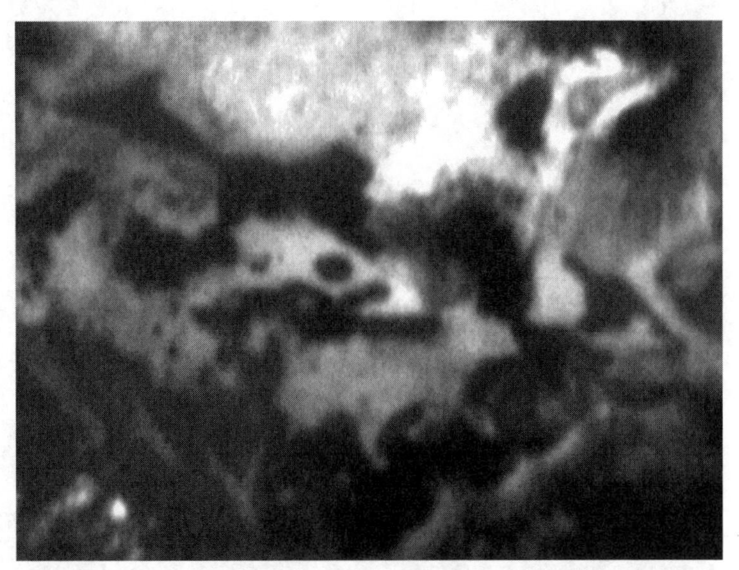

SM70. Geometric shapes (300m) — MOC/MGS Orbiter
Region: South East Floor, West Candor Chasma, Jan 1998
Image credit: NASA/JPL/MSSS

There is all manner of strange stuff going on here which invites caution. However, once again, it is the intriguing geometry at this location, which proffers it up as a possible candidate for evidence of civilization on Mars. Although this time, instead of a preponderance of grid-like rectilinear structure as with our previous town evidence, we now have what appears to be a fascinating array of geometric *shapes*—a diamond shape with a hole in it, sitting on top of a triangular platform (SM70).

But do these shapes exist for real on the surface of Mars and at this location? Is this truly what we are seeing?

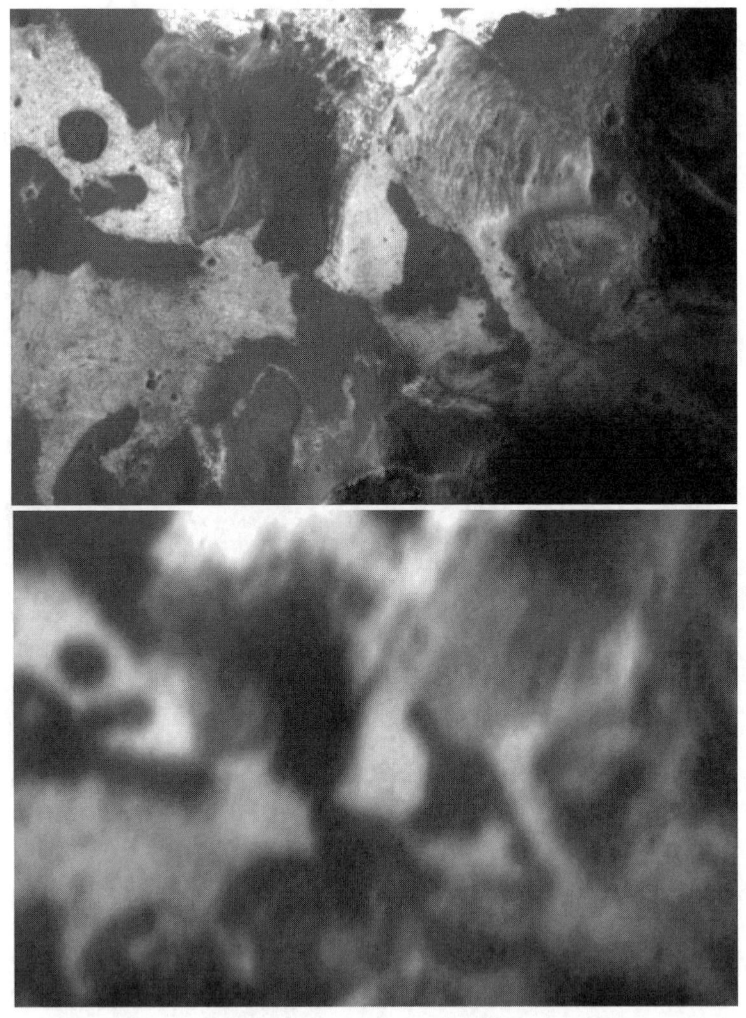

SM70.1 Geometric shapes — comparison of HiRISE (top) and MGS (below) imagery
Image credit (HiRISE, top): NASA/JPL/University of Arizona

In the first edition of this book, I presented the MGS image (SM70) as possible evidence of Martian civilization. This was because I was impressed by the geometry I could see, although I was concerned about the blurred condition of the image. And indeed, when researcher Brian Hopjins recently put me in touch with a HiRISE image of this area, the superior image resolution of 25cm per pixel, revealed a completely different character to the landscape than what the MGS image was able to give us (SM70.1).

And as we can now see, the geometric shapes are not real structures at all but are merely ground level illusions.

This is an investigation that assesses the evidence. Hence, due to the detail revealed in the HiRISE image, I am now retracting this location on Mars as possible evidence of civilization. This is how we must progress if we want to arrive at the truth.

Artificial Objects at Cydonia

Now we will consider some evidence that for many will be familiar: the *Cydonia Mensae* region within which lies The Face on Mars.

My first introduction to the possibility that an ancient civilization once lived on the Red Planet came when I watched a video by Richard Hoagland called *The U.N. Briefing: The Terrestrial Connection.*[64] (Also called "Hoagland's Mars, Volume 2" in the U.S.) This was a presentation given to the United Nations in 1992 in which he gave evidence for the existence of artificial structures at the region on Mars called Cydonia; discussed what benefits this knowledge could offer humanity; and offered his opinion that NASA and others were trying to keep it all secret.

I find it incredible that this information about the Face on Mars and the other objects at Cydonia, has been around now for over thirty years and yet, NASA and the science community have still not engaged with the subject or taken it at all seriously.

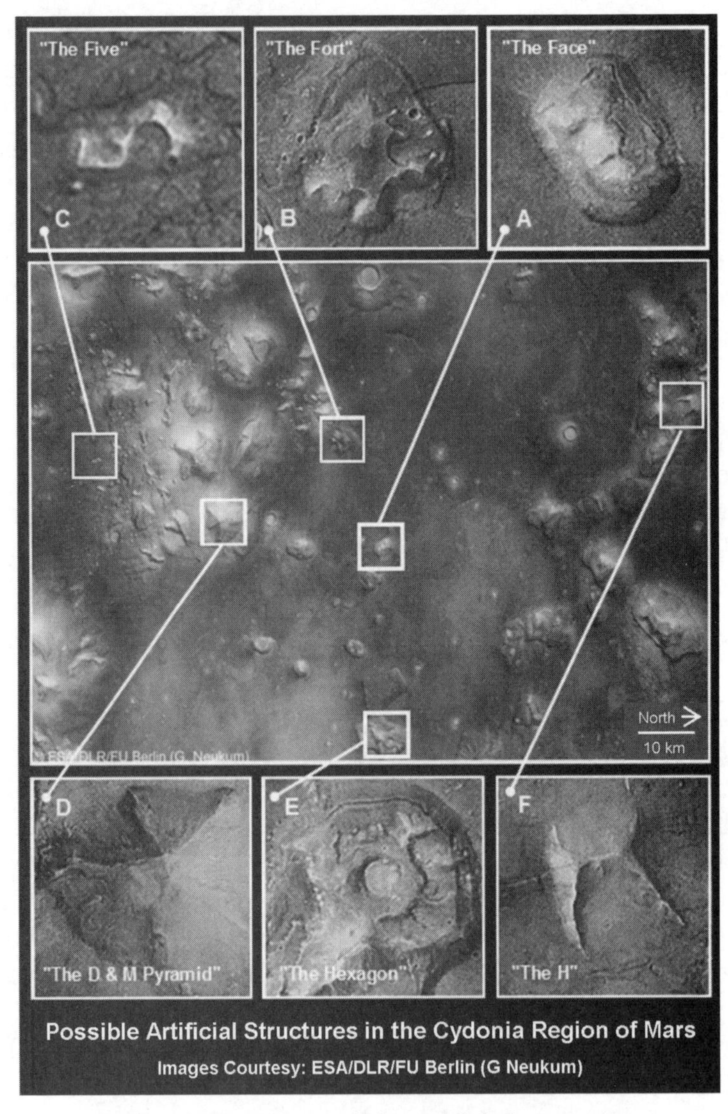

SM71. Cydonia: possible artificial structures
Source image credit: ESA/DLR/FU Berlin (G. Neukum)
(ESA disclaimer: see note[65]); graphics: SecretMars

Scientists have said that because the early Cydonia research was all based upon the low-resolution photos taken by the Viking spacecraft back in 1976, that this research has all been discredited by the new images taken since 1998 by much more powerful orbiting cameras—new pictures of Cydonia that offer a much higher quality resolution and detail than Viking was able to. The main contention therefore of the scientific view being, that the so-called artificial objects such as The Face have been proven by the new imagery to be nothing more than normal geological features such as hills, knobs, buttes, mesas, and massifs.

Well, let's take a look at the Cydonia area now for ourselves and see if we can find any signs of artificiality, intelligent design, and construction.

Cydonia Object A: The Face on Mars

In the first chapter, we presented an image of the Face on Mars that was taken by the Viking Mission back in 1976—an image from which a hypothesis was formulated by several independent scientists and researchers, that the Face was possibly an artificial object.

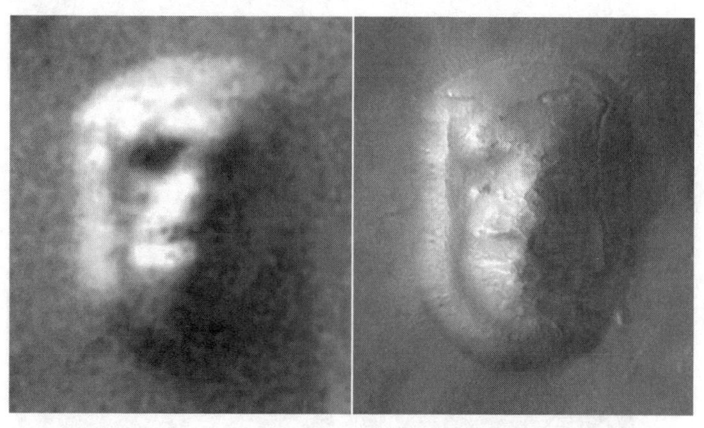

SM72. The Face on Mars: Viking/HiRISE comparison (L) Viking Orbiter 1976 (R) HiRISE/MRO 2007
Image credits: (L) NASA/JPL (R) NASA/JPL/University of Arizona

In 1998 and after much public pressure, NASA finally decided that new, more detailed pictures were to be taken of the Face, to satisfy researchers and the public and ostensibly to settle once and for all the argument as to whether or not it was an artificial structure. These pictures were taken first by the MOC camera of Mars Global Surveyor in 1998 and 2001, and then a few years later in 2007, via the HiRISE camera of Mars Reconnaissance Orbiter. Also, the European *Mars Express* obtained some lesser detailed images in color.

For our analysis, we will use the 2001 image from Mars Global Surveyor, which was taken at a resolution of two meters per pixel (SM73).

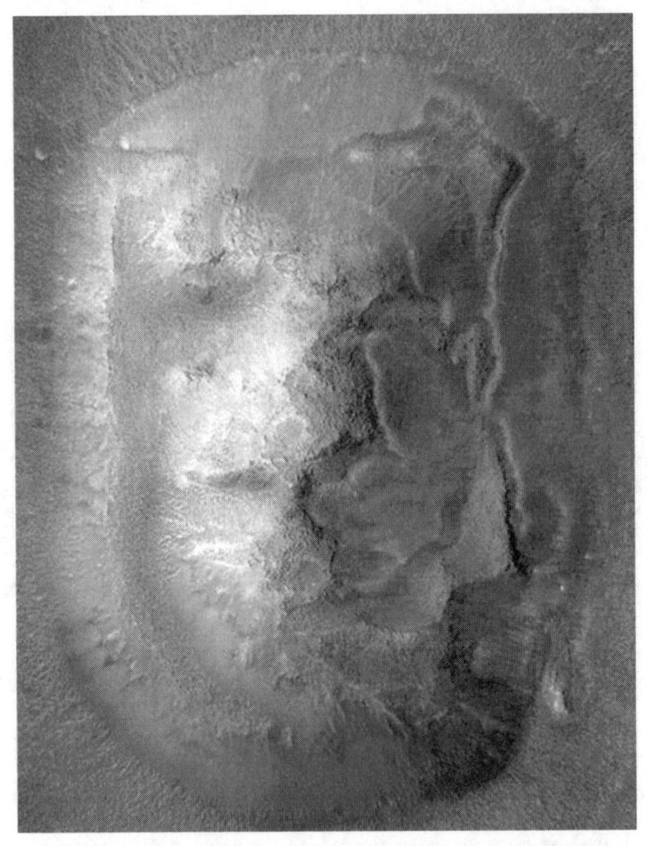

SM73. The Face on Mars: 2001 (2.75 x 2 km) — MOC/MGS Orbiter
Region: Cydonia Mensae, Apr 2001
Image credit: NASA/JPL/MSSS

The mainstream scientific opinion of the Face on Mars is that it is nothing more than an eroded mesa, a remnant massif, probably formed through landslides and an early form of debris apron. The view of researchers supporting the artificiality hypothesis, however, is that it is possibly an artificial monument carved or engineered from the rock.

So what can we see that might persuade us we are looking at an artificial construct instead of a hill that just looks like a face?

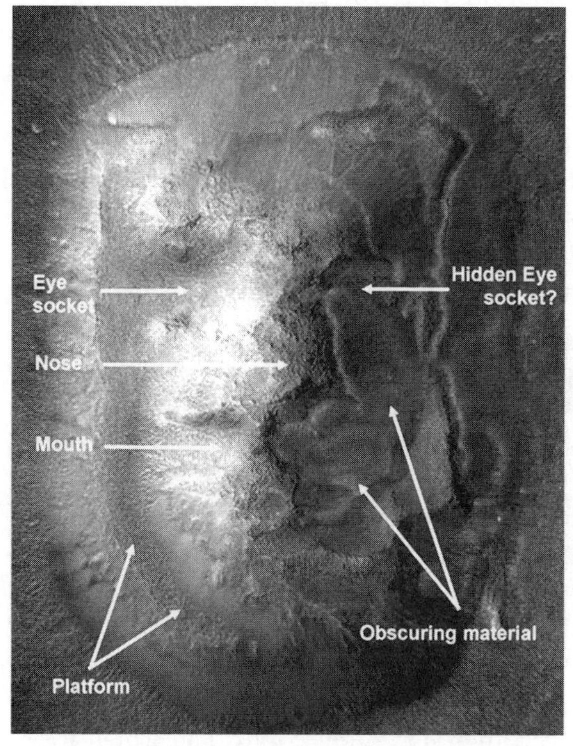

SM73.1 The Face on Mars: 2001 (detail)

The one thing that strikes me first about this object is the overall shape of it; a harmonious curvature and symmetry that is quite elegant to the eye. Also, there appears to be a level platform upon which the central features of the face are sitting, which could be indicative of a design element.

The left side also is a pretty good shot at depicting a face, with an eye socket, nose, mouth, and forehead all proportionately accurate. However, for this object to look like a complete face, what we need is for those details to be mirrored on the *right-hand side*. We would then have little doubt that this was a monument created by a civilized race of intelligent people. However, what we have got is a conundrum.

Unfortunately, there is an accumulation of material on the right side of the face which is hiding whatever might be beneath it, right where we need to find a corresponding eye socket and extension of the mouth. So, unless we can sweep away all that material or somehow scan and penetrate underneath to reveal what's hidden there, we are not going to be able to settle the question as to whether the Face on Mars is artificial or not. As such, I don't believe we are going to gain much more insight from studying the actual image further.

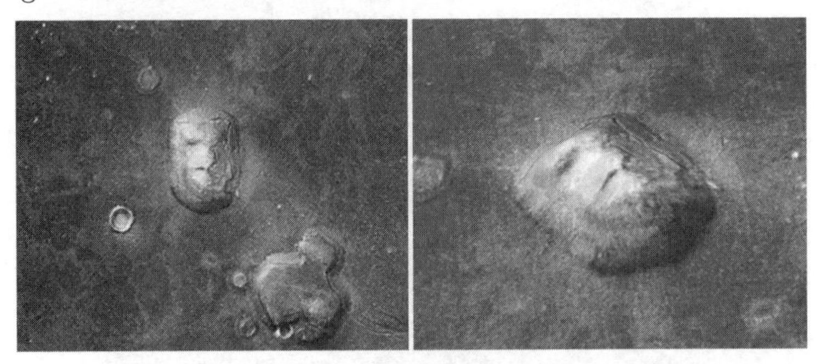

SM74-75. The Face on Mars: 2006 — Mars Express
Image credits: ESA/DLR/FU Berlin (G. Neukum)

However, there are other contexts in which we can evaluate whether the Face is possibly artificial or not. For example, when we get up close to it the Face looks just like an eroded, weathered bunch of rocks, like any other place, but if some Martians had decided to create this mega-sized monument—and it *is* a mile and a half long and over 1,000 feet high—we might consider what their intended use for it was. For example, was its function to be seen and recognized as a face only from *high altitude?* If so, then they have certainly achieved that goal already (SM74-75):

Also, we need to consider the effect that planetary devastation would have had on any buildings or monuments standing in the path of raging tsunamis, destructive blasts of incinerating heat, and massive meteorite bombardment. Also, the many thousands of years or more, of aging and erosion on top of that. Needless to say under such conditions, to find an intact, large-scale building without blemish and looking in prime condition would be completely and utterly unrealistic.

In which case the right side of the Face that we see obscured by material, may have been the side that met the impact of blast waves. For, if we look at the whole area of Cydonia there is a pattern that seems to show a large number of hills, mountains and features with a pile of material heaped up on one side only and facing the same direction—the same side on which the Face has its deposited material.

Interestingly, Cydonia lies on the dichotomy that separates the Martian highlands from the northern lowlands where, it is theorized, a great impact stripped away a huge chunk from Mars, blasting some of it off into space and heaving the rest back onto the other side of the planet in a nightmarish hail of destruction. Significantly, the northern lowlands lay an average of 4-5 kilometers lower than the rest of the planet.

There is, however, a more conventional geological explanation for this accumulation of material and it's called "Pasted-on-terrain", a covering mantle of dust-covered snow and ice. As this material is deposited on the side of the Face that receives the least amount of exposure from the sun this could explain why it hasn't eroded and melted yet like the other half.

There are other theories too for the obscured right side, which include Hoagland's idea that this was an engineered monument originally covered by perhaps metal and glass, and that what we see now are the effects of the blasting away, erosion or melting of that original covering by the calamitous events that struck here eons ago.[66]

So, we've now seen some detailed pictures of the Face on Mars and considered the context in which this object may have been built by ancient Martians. What do you think? Is it an artificial object or just an ordinary geological feature that happens to have some similarities to a humanoid face? Well, taken by itself and in isolation, overall I would not consider the Face on Mars to be an artificial construction. There are just not enough elements of potential artificial design that could not be achieved by geology.

The interesting thing about Cydonia however, is that several other unusual objects in this region also display possible artificial signatures. Therefore, if we can find more than one such object in the same area as the Face, it would provide us with corroborative evidence that it just might be artificial after all.

So let's take a closer look at some of these other strange features that lay just several miles away from the Face on Mars…

More Artificial Structures at Cydonia?

Cydonia Mensae lies in the northern hemisphere at approximately 40.2° north and 350.2° east and is set on the boundary between the cratered highlands of western Arabia Terra and the flat northern lowlands of Acidalia. Geologically speaking, Cydonia is regarded as a transition zone that features impact craters and eroded mesas, wide expanses of smooth ground, and few signs of volcanic activity. It's a strikingly interesting area with a variety of strange shapes and odd-looking features referred to as knobs, buttes, and massifs. From a geological and therefore scientific perspective any pyramidal or artificial-looking features found in this area must have been created by natural processes of erosion on Mars, and there is nothing unusual about them at all.

Let's see if this is an accurate statement.

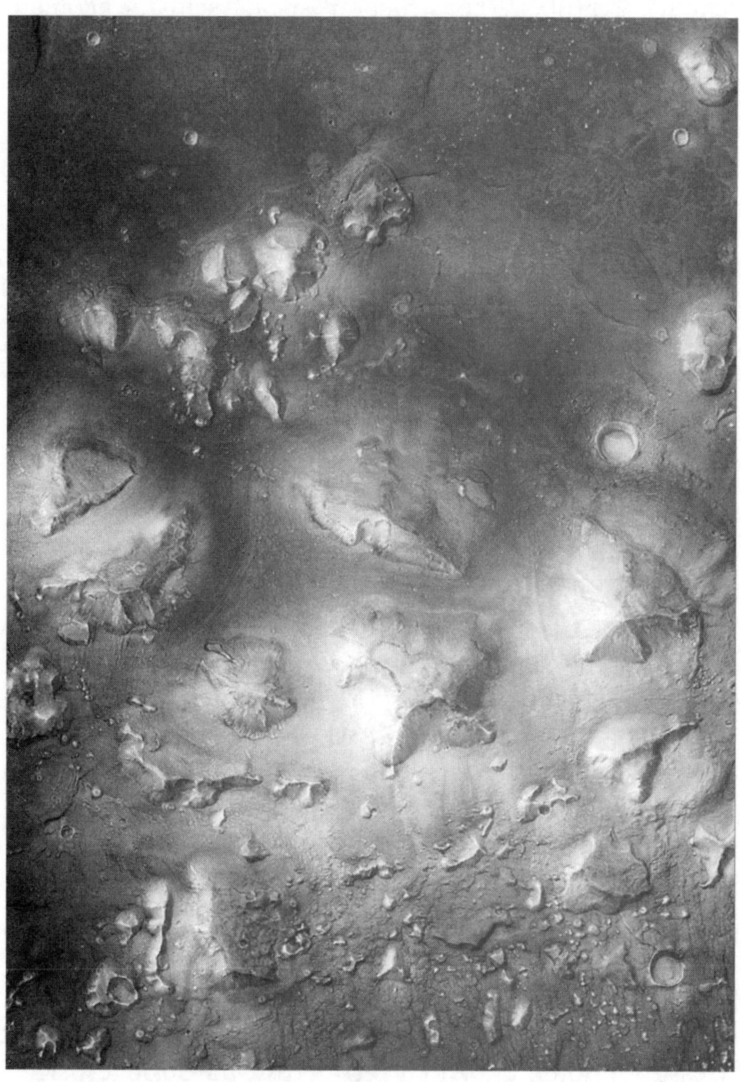

SM76. The Cydonia Region: 2006 — Mars Express
Image credit: ESA/DLR/FU Berlin (G. Neukum)

There are many anomalous features and landforms in the Cydonia region, and I've chosen to present as possible evidence of an ancient Martian civilization only those that support most of the criteria upon which we are pinning our search for artificial construction: strong signs of geometric form.

Cydonia Object B: The Fort

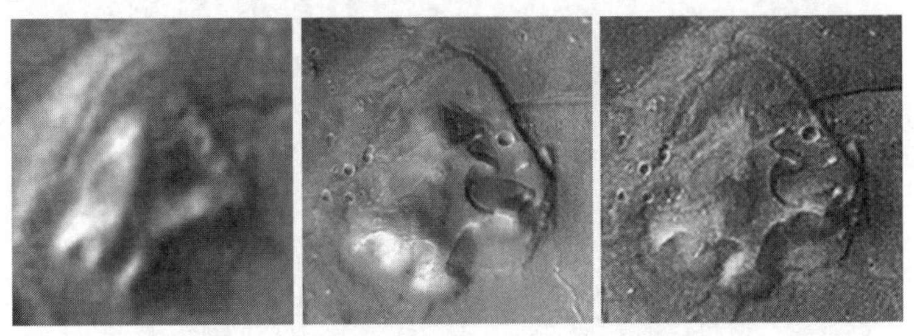

SM77. The Fort: comparison 1976-2006
(L) Viking Orbiter (Centre) MGS Orbiter (R) Mars Express
Found by R. C. Hoagland
Image credits: (L) NASA/JPL (C) (i) NASA/JPL/MSSS
(ii) Dr. Mark J. Carlotto (R) ESA/DLR/FU Berlin (G. Neukum)

We are now going to look at what is called the "Fort", a name chosen by Hoagland mainly, I think, because, in the original Viking image, shadows depicted a trapezoidal form with a triangular enclosure (SM77L). One can easily see after viewing the Viking image why this feature was thought to be artificial back then. The triangular shape is just striking. Of course, now that more detail has been revealed by the better cameras, we can see quite a different character to its shape, even though it still gives us cause to wonder about its origin.

Let's take a closer look. In this image of the Fort (SM77.1), we show a pattern of linear and geometric consistency that may be revealing signs of an underlying artificial structure. Measuring some 2.8 kilometers across, the strongest sign of an inherent artificial element to this formation is the basic platform which has a rounded triangular shape to the north and a sharp, right-angled edge to the east. There is also the suggestion that a more elaborate construction sits on top of the platform, one that includes the curved features to the south-west and eastern side, but as these are somewhat less obvious we have not outlined them here.

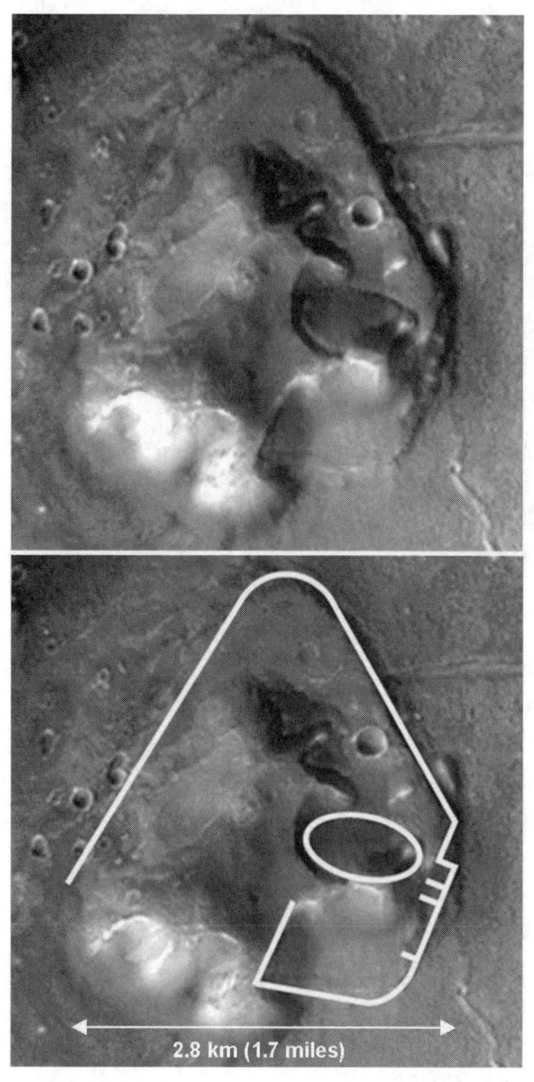

SM77.1 The Fort: geometrical pattern (2.8 km/1.7 miles)

There do appear to be several elements of artificial design to the Fort and an overall architectural harmony that gives one the feeling we are indeed looking at something intelligently laid out. It's as if a mystery won't quite reveal itself here, but you sense something is going on. Or, at least it did a very long time ago. The edges are softened, eroded, maybe damaged even, and the structure covered by many layers of soil and sediment perhaps several thousands of years old.

Cydonia Object C: The "Five"

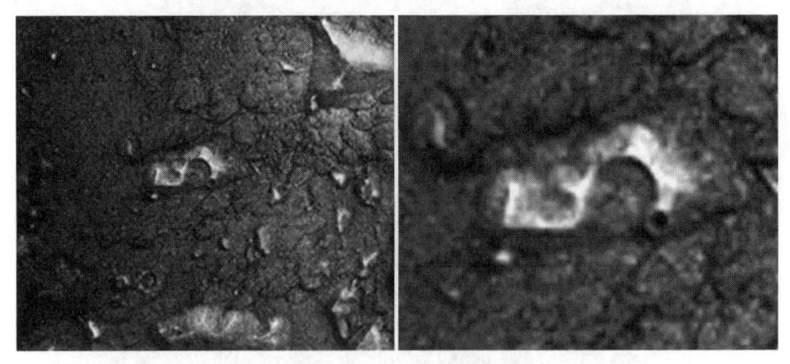

SM78. The Five (1.2 km) — Mars Express 2006
Noted by M. J. Craig; image credit: ESA/DLR/
FU Berlin (G. Neukum)

The next mysterious feature we're going to look at is something I haven't seen mentioned too often by other researchers but I believe it's certainly worthy of our attention (SM78). I've called it the 'Five' because if you rotate it 90 degrees to the right…

Anyway, for illustration purposes, I don't want to show it that way around because it will otherwise distract us from the amazing angular definition of this object. I think the shape is extraordinary, and if what we are seeing is truly an accurate reflection of what exists on the ground, then this surely must defy a geological explanation, although I'm sure someone will have a go.

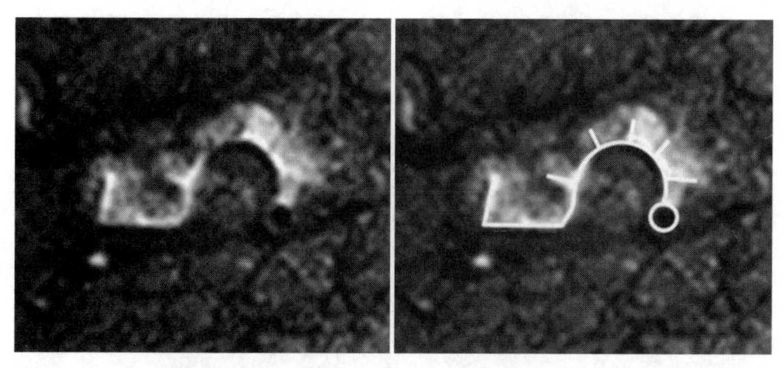

SM78.1 The Five: geometrical pattern

There are **four** powerful aspects to this object that persuade me to lean towards an artificial explanation:

1. **The precise right-angled turn** located on the far left of this formation.
2. **The half-circle** and the suggestion of equidistant 'spokes' emanating from it.
3. **The perfectly aligned connection** of the half-circle and spokes, with a small round object to the far right.
4. **The constant *width*** maintained that runs consistently across the whole thing, which I estimate to be about 50 meters. The length of the entire object is about 1.2 km.

The one reservation I have about this picture is that it is only some 15 meters per pixel resolution, and when you start to magnify these images, data compression artifacts and the camera limitations begin to come into play in that digital pixel patterns start to interfere with the accuracy of what you are seeing.

Without going into the technical detail—not my specialty what is happening is that at this level of zooming-in you start to see a pattern of interlaced detail that tends to create 'straight lines and blocks' out of something that is possibly not entirely straight at all. This is something we must keep in mind as we study these images, especially when we are attempting to extract detail from a lower-resolution picture such as this one.

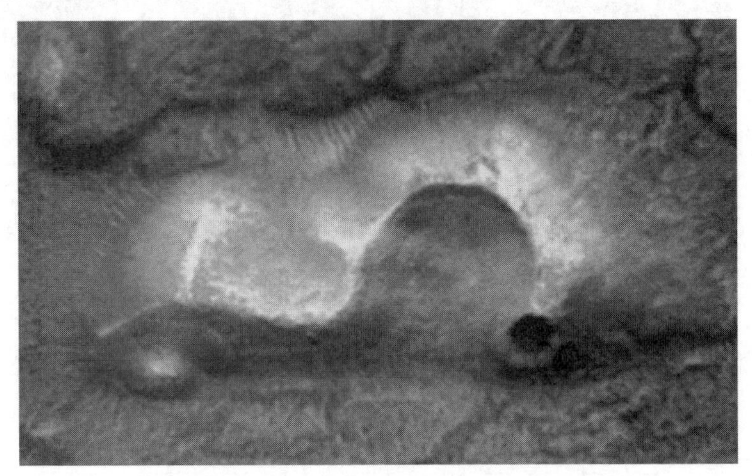

SM79. The Five — HiRISE/MRO 2008
Image credit: NASA/JPL/University of Arizona

For this reason, I went in search of an image that hopefully would give us more detail to try and confirm the existence of this remarkable object.

Mars Global Surveyor took several pictures of the area we are interested in, capturing several long, narrow strips of image data, and I did find one that managed to snap a part of it. But the best image was found by researcher George J Haas, who kindly sent me a link to a picture taken by the HiRISE camera that shows the entire structure (SM79).

As we can see, this new image certainly confirms all of the straight lines and angles—except perhaps for the radial spokes—and even detail of yet another straight edge coming off from the corner of the right angle.

As we now have confirmation that this object does indeed exist and that it's not an illusion created by low-resolution image artifacts, I consider it to be potentially very strong evidence of artificial construction. What the structure may be, of course, is entirely open to speculation . . . perhaps the site of an ancient library . . . now wouldn't that be something worth exploring!

I just find this structure captivating. It fires all of the geometric signal flares we are looking for, providing as clear a sign as any that we must surely be looking at an artificial structure here. Or at least the remains of one perhaps partially buried, damaged, or in whatever condition.

Cydonia Object D: The D&M Pyramid

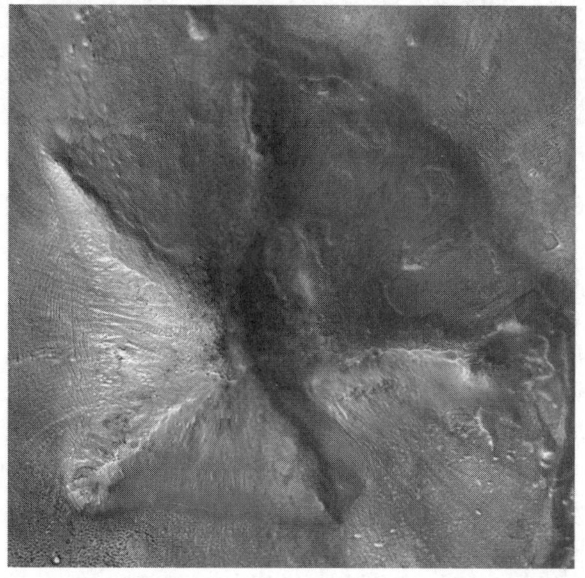

SM80. The D&M Pyramid (3 km) — MGS/THEMIS
Found by V. DiPietro/G. Molenaar
Image credit: NASA/JPL/MSSS

Now we come to the "D&M Pyramid", named after its discoverers, scientists Vincent DiPietro and Greg Molenaar. Standing over a kilometer high and about three across, this massive formation could well be a five-sided pyramid.

This object is normally viewed from south to north, displaying an emphasis on the three fairly intact western sides (SM80), but by rotating it instead from north to south, we get to appreciate the remarkable star-shaped structure much more clearly, showing its very symmetrically proportionate, protruding limbs (SM80.1).

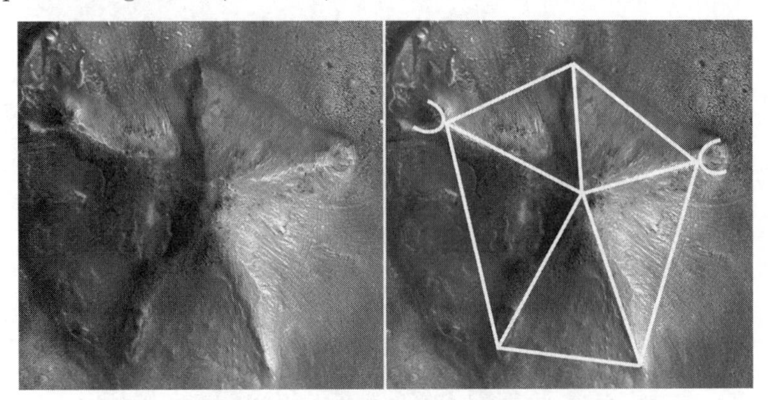

SM80.1 The D&M Pyramid: geometric perspective

I think this is a more remarkable object than the Face especially in terms of its striking symmetry. The likelihood of a geological process throwing up this five-faceted formation is surely remote.

It's an intriguing prospect to imagine what wonders may lay within such an enormous structure, and it's a shame on NASA they are not even curious about it. Are we are beginning to see just a few too many of these geometric shapes in the same locality? This object is only 20 km southwest of the Face; it is also just 15 km from both the Five and the Fort.

If we look at the map and diagram where these formations are located (SM81), we can see what appear to be many other mountains and mesas, and it is because there are so many of them that geologists rationalize you are bound to find a few that might give the appearance of being odd and artificial-looking, which is a fair point. However, even though my ignorance of geomorphology will possibly rear its head here, I will nonetheless reserve the judgment of what my instinct tells me, which is, that there's much more to this region than meets the eye and we should all prepare for a surprise.

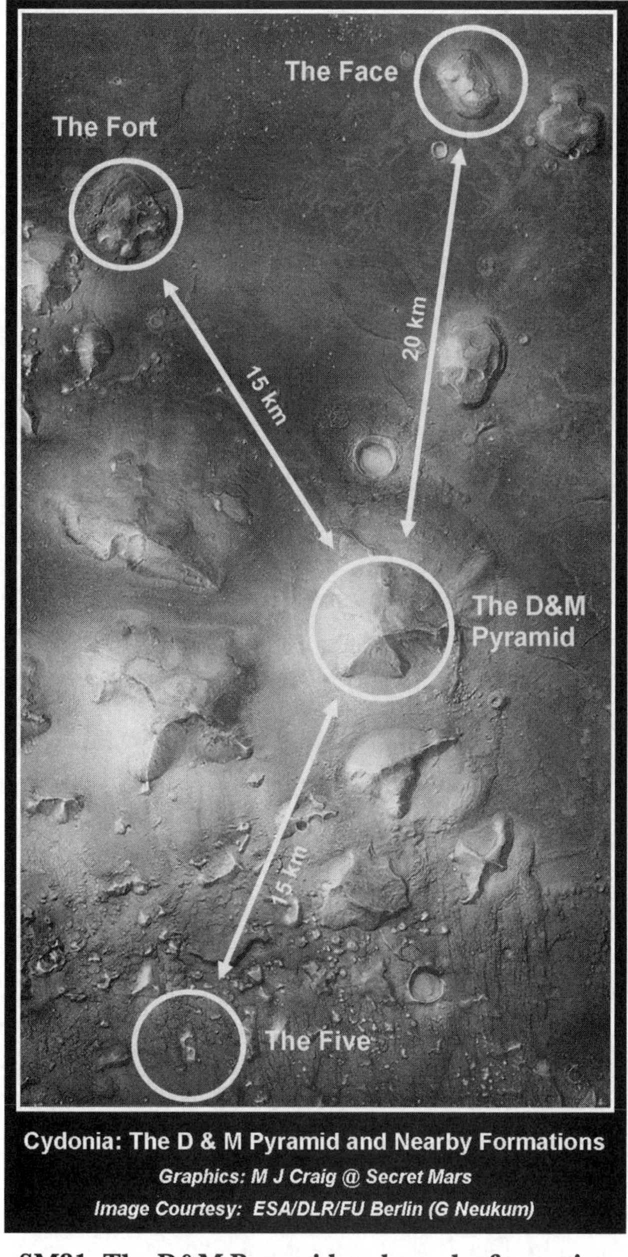

Cydonia: The D & M Pyramid and Nearby Formations
Graphics: M J Craig @ Secret Mars
Image Courtesy: ESA/DLR/FU Berlin (G Neukum)

SM81. The D&M Pyramid and nearby formations

Cydonia Object E: The "Hexagon"

SM82. The Hexagon (2 km) — Mars Express 2006
Image credit: ESA/DLR/FU Berlin (G. Neukum)

Here is another image in Cydonia that I believe should warrant some attention (SM82).

What caught my eye about this one is the hexagonal shape that appears to sit within another geometric shape. There appears also to be some heavy erosion or impact damage to the western edge, which makes it difficult to see what the original formation may have looked like if indeed there was an original shape to it.

Whenever we see these precise angles and turns, as with this object, we surely have to question why and how this can happen. The fact is that hexagonal structures in nature are quite common as we see them both in ice and rock formed by a variety of processes. However, when we see geometrical precision displayed on such a massive scale as this, and particularly when found in proximity to other geometrically proportionate formations, it is surely time we considered other explanations than geology.

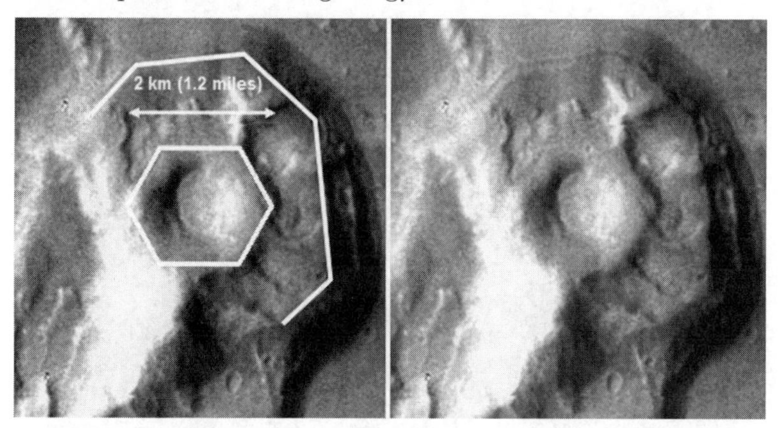

SM83. The Hexagon (2 km) —THEMIS/Mars Odyssey 2002
Source image credit: NASA/JPL/Arizona State University
Graphics: SecretMars

I found another image that at least confirms the pattern although it doesn't give us any more detail (SM83). I've outlined the basic hexagonal shape and the fairly straight edges of the outer formation, which I think are very interesting and which might be betraying an underlying artificial complex.

However, there is another idea that might explain this formation, which is that of *electrical arcing* upon the planet's surface.[67] There is very strong evidence that many craters are formed not by meteorite impacts, but instead, by huge bolts of electricity. Laboratory experiments by plasma physicist C. J. Ranson have shown that hexagonal features can be created by electrical arcs, and as there have been several hexagonal craters found on Mars and throughout the solar system, I think this possibility should also be considered for this particular site.

Cydonia Object F: The "H" Structure

SM84. The H Structure (2.5km) — Mars Express 2006
Image credit: ESA/DLR/FU Berlin (G. Neukum)

This next object is quite amazing (SM84). The parallel symmetry is visually impressive and, if I may venture to say, architecturally elegant. Geologists will insist that what we are looking at is merely a remnant landform of hard rock that has remained after all the softer rock surrounding it has become eroded over millions of years, which is pretty much the scientific explanation given for most of the formations we have presented from Cydonia.

But why should such an *unusual shape* come to exist after all that time and erosion, and with such smooth symmetry? And if it is presumed to be just a random occurrence in nature, then why are we finding so many 'random formations' in this region that possess geometrically consistent features?

Here's an overview of other features in proximity to the H Structure (SM84.1):

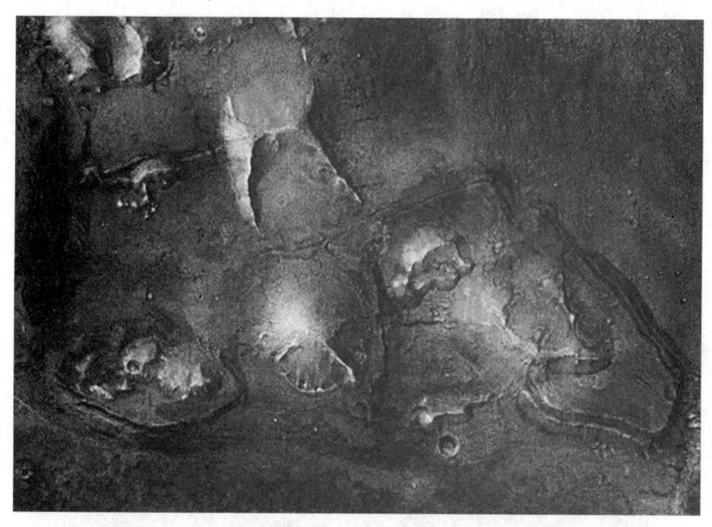

SM84.1 The H Structure and nearby features

When I look at this picture I just get the sense that we are once again looking at an intelligent layout. There just seems to be something in the spatial relationships between each of these objects that suggests deliberate placement. They are not bunched up awkwardly against each other, as you might expect in random geology, instead, there is comfortable space in between every feature. This I find intriguing.

NASA geologists declare that there are thousands of these knobs, buttes, and mesas spread out across Cydonia and that they are nothing special. That may well be for most of them, but the more relevant point I think is why do we find so many geometrically interesting features in this particular region?

Some research has suggested that these features sit on the coastline of what was once an ancient Martian sea. Are we then perhaps looking at several clusters of building complexes, maybe communities spread out over this region of Cydonia that used to live by the coast?

Well, there are many more features and objects that I could include here from the Cydonia region, but these will do for now. This means we've pretty much wrapped up the visual image evidence of Mars that I wanted to show you in this book.

Summary of Evidence: Signs of Civilization on Mars

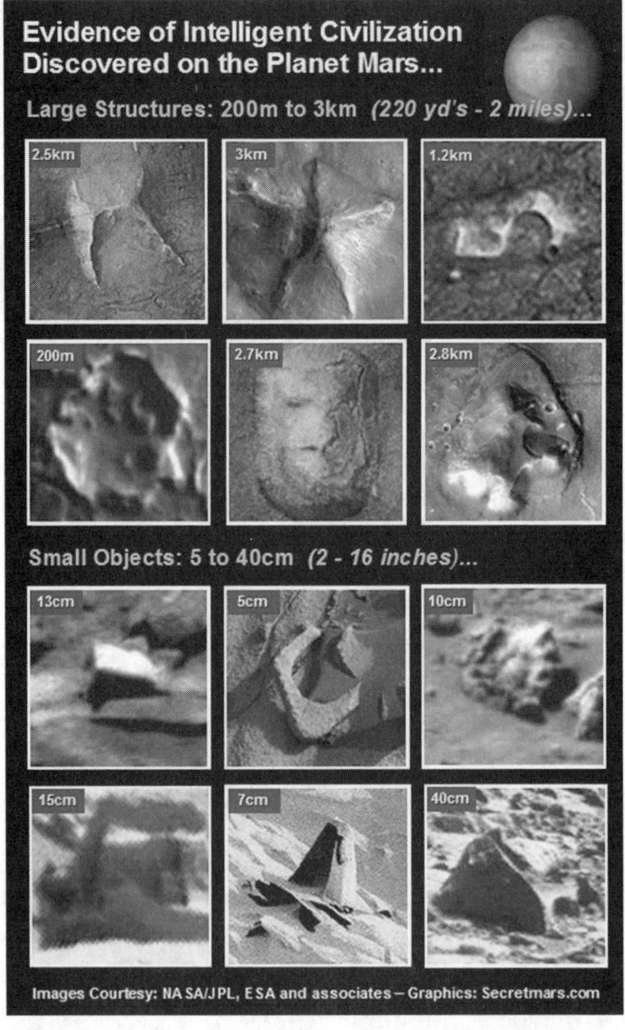

SM85. Evidence of intelligent civilization on Mars

Have we found sufficient proof that an ancient civilization once lived on the planet Mars? Well, in my opinion, and considering each picture *in isolation by itself* of both large-scale evidence and the much smaller objects we've seen, I would say "No" we do not have any single proof. Virtually every picture could be accounted for by a geological explanation or a mundane one such as an image defect or illusion caused by perspective, shadow, or low image resolution.

However, considering all the images *together as a broad spectrum of evidence*, I would say "Yes", we do have enough proof. There are far too many artificial-looking objects and geometrically consistent formations to be easily dismissed as random occurrences in nature.

What persuades me towards this conclusion is the sheer simplicity, yet clearly defined structure, of some of the smaller objects we've discovered, and too, the locations where they were found:

1. The circular cog mechanism, with the regularly-spaced protrusions.

2. The cube-shaped box.

3. The crank handle object.

4. The many other bits and pieces of seemingly artificial debris.

The three main rovers Spirit, Opportunity and Curiosity, have traversed and explored only three locations on Mars, and although Opportunity seems to have turned up far less of these objects at Meridiani Planum than Spirit and Curiosity did at Gusev and Gale Crater, it is far too low a sample rate to be able to assess a true picture of the possible extent of ancient sites of civilization on Mars, at least with regards to small-scale evidence. Therefore we can't argue yet that it is a telling factor that more objects were found at Gusev and Gale, although it could be.

However, we can say that many of the large, geometrically proportioned formations were indeed concentrated in the Cydonia region, and a similar concentration of such large objects has so far not been identified, although analysis by independent researchers remains ongoing.

A concentration of evidence in several locations on Mars, would, of course, give us a firm, rational and objective basis for our hypothesis but we don't quite have enough evidence of that yet. Nonetheless, I'm prepared to accept the evidence presented so far for exactly what it appears to be: signs of an intelligent civilization having once existed on Mars. That is my conclusion. Whether you the reader have seen enough evidence to accept this view, I will leave to your good judgment.

NASA Website to Explain Anomalous Objects

NASA scientists could prove that none of the controversial images in this book are unusual at all and that every one of them can be easily explained. If convincing I will of course humbly accept their evidence.

However, they could already have helped to avoid any suspicion and doubt as to their motives by providing a simple public service web page: an official NASA source to identify and explain every unusual object imaged by their orbiting spacecraft and rovers, which might cause the public to question what it is.

In one stroke they could perhaps make books such as this one redundant. Provided we are given the truth.

When Did Mars Have a Breathable Atmosphere?

On balance, my opinion is that we have found evidence of a past civilization. But there is quite a serious issue with accepting this conclusion—and that has to do with the planet itself.

Here's the problem. The atmosphere on Mars is 95% carbon dioxide, absolutely unbreathable for humans. If we are looking at evidence of a Martian civilization and presuming they were similar beings to us that needed to breathe the kind of air that we do, we have to ask the question: how did that civilization manage to survive on the planet's surface, if Mars has been a dead world for well over a billion years or more? And, crucially, when the possible ruins we are seeing in NASA images can't surely be any older than a million years at most?

This is potentially a huge problem for validating our evidence, as the time differential is enormous. The crucial question therefore is: *when* did Mars have a breathable atmosphere, and *when* did it disappear?

Most scientists believe that the planet Mars we see today has been this way for a very long time, meaning it has had no liquid water flowing on its surface or breathable air for millions of years, if not much longer. This would logically suggest that if there had been a civilization on Mars, it could only have existed all those millions of years ago when the planet was theoretically capable of sustaining life. The problem, therefore, is pinpointing exactly when Mars became a barren planet and caused any Martian civilization in existence to meet its demise.

Which leaves us with the perplexing question: presuming that the objects in Cydonia, for example, are indeed the ancient remains of a past civilization, how could we be looking at ruins that are *many millions* of years old? Surely it would be reasonable to assume that erosion would have reduced them to dust over such an enormous stretch of time.

However, erosive conditions may not have been that severe, because after the cataclysm that struck Mars there would have been no rain, rivers, or moisture to accelerate the process of natural erosion. But even still, I find it hard to reconcile the time difference between the era when Mars may have had air to breathe and thus been able to sustain a civilization of living beings, to when those conditions disappeared right up until today when we are now supposedly still able to see the remains of that civilization's buildings in the Martian soil.

I find it a struggle to accept that we could hope to recognize archaeological remains, even after 100,000 years of erosion, let alone over a million years or more.

We are presuming here of course that the people who built the civilization were walking around, breathing, in a stable, healthy world. For this to be possible Mars must, therefore, have been a living planet for much longer than science is telling us—up to at least a million years ago or so. But this is difficult to rationalize, as the scientific studies of the planet tell us that Mars has been a dead world for way longer than that.

There is, of course, the possibility that the planetary scientists are wrong about their assessment of when Mars died. Or even how early in its history, Mars became capable of evolving or sustaining life. The currently valid scientific evidence supports their conclusions, but maybe all the important facts concerning Mars' planetary evolution are not in yet, and these crucial events we are speaking of may have occurred much more recently in Martian history.

If Mars gradually lost its life-sustaining atmosphere over a long and extended period, rather than suddenly, this may have given a civilization time to adapt to the new conditions, and, if they were losing breathable air, to perhaps enclose their towns and cities within protective domes. Or even relocate their civilization underground instead of on the surface.

And there is also one other possibility to consider, one which may make the whole preceding argument somewhat redundant.

Chapter 5: The Alien Connection

We have been assuming that the existence of a civilization on Mars would be entirely dependent on that planet's ability to provide a living habitat; a world capable of providing air, water, and an atmosphere to sustain its living creatures and inhabitants; a protective world that had been gradually produced through eons of geological time, where indigenous Martians could go about their daily business of building their homes and monuments on warm, sunny days while breathing the fresh Martian air.

We would also be assuming that this race of sentient beings would have appeared by way of a similar evolutionary course to what we have been taught gave birth to the intelligent species found on Earth: primordial soup ... lightning ... primitive life ... complex life ... apes ... humans ... trees ... caves ... settlements ... villages ... and civilization.

But what if this scenario is one built on a false premise? What if this whole story of evolution producing an intelligent race of thinking beings that arose from the primordial slime four billion years ago is an untypical myth for the appearance of life on worlds throughout the galaxy? What if intelligent life came to Mars and indeed the Earth in quite a different way?

Colonization of Habitable Worlds

If we have found evidence of an intelligent civilization on Mars then this immediately throws up a massive question, which is: were they just a primitive race that built stone monuments, or, were they technologically advanced and had achieved space flight?

It is a question that blasts open a huge can of worms. For, if a discovery was indeed made by the space agencies on Mars which proved beyond doubt that a technologically advanced race once lived there, then such a revelation would have given them a massive cause for concern in wanting to keep that information quiet.

We can just imagine the uproar if religions which believe that 'God's plan' was to only create intelligent life here on the Earth, are told that buildings made by intelligent beings have been found on a different planet in our solar system. And that those beings were once able to travel into space and very likely visited the Earth—perhaps even seeded life here.

Where would that leave our scientists too, who've been teaching for well over a century that human beings are indigenous to the Earth and have evolved from apes? How confusing for them to then have to consider the possibility that most intelligent life in our galaxy may not arise indigenously on a planet through evolution, but instead spreads rapidly from the activity of just a few civilizations who are intent on exploring space . . . bold adventurers who would seek out and find all life-sustaining worlds like the Earth, and then proceed to *visit and colonize them.*

Just as we are now, at this very moment, searching for and logging Earth-like worlds through our giant telescopes with the Kepler Mission, civilizations just a thousand years older than ours and way more advanced technologically and in their ability to travel to the stars may have done this already many centuries ago.

For if we, the human race, in our space-faring infancy, are just now making plans for colonizing the planet Mars, what is the chance that a much more advanced race has already been there and done it, on Mars and many other planets spread throughout our vast portion of the galaxy; hundreds, perhaps many thousands of worlds, all now having a common ancestry?

And when they decided upon a world to inhabit they would bring with them most of the things they would need to set up home and civilization, just like the colonists on Earth did when they sailed the Atlantic Ocean to the New World four centuries ago: including the means and technology to construct their homes and facilities . . . probably enclosed in protective structures isolated from the harmful local atmosphere.

The appearance therefore of a civilization on Mars would no longer be dependent upon whether or not the planet had breathable air or even the conditions for the evolution of life. All that would be required for intelligent life to appear on Mars is for a space-faring civilization to decide to target the Red Planet for colonization. In fact, exactly as we are doing now, as NASA contemplates colonizing a world where *human beings cannot breathe without a spacesuit and where they will need to live in protective habitats. . .*

Science and Extraterrestrial Visitation

The fact that human beings still using primitive rocket propulsion are planning to colonize Mars, surely underpins the hypothesis that intelligent races in our galaxy far more advanced than us and with far better technology could have done so millennia ago. Therefore, rather than regard such a possibility as extreme and fanciful as some would still argue, I believe we should consider this idea instead, to be an entirely reasonable and rational one.

Judging by the official documents now being released by the governments of Argentina, Australia, Brazil, China, Denmark, Finland, France, Germany, India, Ireland, Japan, Mexico, New Zealand, Peru, Russia, Spain, Sweden, Ukraine, Uruguay, Vatican City and by some departments in the United Kingdom and the USA, the case file for extraterrestrial craft visiting the Earth which has been a highly classified subject for decades, is an enormous one. Even the CNES (The French National Centre for Space Studies), asserted the absolute seriousness of UFO research through a document and study released in 1999 called the COMETA Report.

Entitled *UFOs and Defence: What Should We Prepare For* [68] this document is an independent report that contains the results of a study made by IHEDN (French Institute for Advanced

National Defence Studies), whose role is to ultimately make recommendations to the French Prime Minister and Head of State to consider. This document was made freely available to the French people, which is the way it should be.

The COMETA Report made this summary:

> The accumulation of well-documented sightings made by credible witnesses forces us to consider from now on all of the hypotheses regarding the origin of unidentified flying objects, or UFOs, and the extraterrestrial hypothesis in particular.
>
> — *Denis Letty, French Air Force General*

With so many studies and testimonies bearing witness to these visiting craft, this observation alone suggests that our galaxy is already populated by several intelligent civilizations that are capable of traveling to other solar systems. Sgt Clifford Stone, at the Disclosure Project Washington Press Club Conference in 2000,[69] stated that to his knowledge, U.S. Army Intelligence had identified, cataloged, and described at least 57 different extraterrestrial races that had visited the Earth.

To travel to our planet, this means they must have the ability to travel faster than the speed of light—much faster in fact if journeys between stars are not to take years. Unfortunately, it is mainly for this reason that mainstream science will not seriously consider the possibility that extraterrestrials may be visiting the Earth and our solar system. Our scientists obstinately refuse to believe that anything can travel faster than light, therefore in their eyes there just *can't be* any aliens visiting the Earth because it would take them years and years to get here, therefore making interstellar travel impractical. "Faster-than-light travel is not possible, therefore aliens cannot be coming here"—end of their equation.

The arrogance with which our scientists can so easily dismiss the notion that there might be other civilizations, thousands of years older than ours, and that may have discovered how to travel faster-than-light and are therefore quite able to visit the Earth seems to me almost infantile in its naivety. Simply beyond their paradigm of thinking, it would seem.

I think it is a case of head-in-the-sand syndrome. To my mind, it seems entirely plausible and rational. I mean, what would we expect from our *own* human civilization a thousand years from now? Will we have solved the problem of achieving faster-than-light travel? I would expect so. Or do we simply assume the mantle of those scientific dons who once proclaimed that *air* travel, followed by *faster-than-sound* travel and then *space* travel, was all impossible until we actually did it?

Now it is *faster-than-light* travel that seems quite impossible. Has our scientific fraternity learned nothing from their forebears? I've heard brilliant physicists utter the most ridiculous statements much to the tune of:

"Well, we've pretty much figured out most of what there is to know about the universe…"

The fact remains however that highly-trained aircraft pilots, air traffic controllers, military personnel, police officers, as well as every kind of person from all walks of life have seen aerial maneuvers impossible for terrestrial craft and witnessed alien-looking vehicles landing on the ground and flying through our skies, without fail, year after year.

Rational analyzes of UFO sightings will of course routinely dismiss 80-90% of cases as mistaken conventional aircraft, weather balloons, Chinese lanterns, natural meteorological phenomenon, or a variety of other valid explanations, with a percentage of the saucer and triangular-shaped craft also likely to be secret military projects either man-made or reverse-engineered from recovered extraterrestrial craft.

But the consistent 10% or so of UFO encounters remain inexplicable, other than in terms of a visitation from extraterrestrial or extra-dimensional, intelligent beings to our planet. It is the most likely and logical explanation and defies anyone who has taken the time to analyze some of the vast evidence that is widely available to the public.

However, despite the ever-mounting evidence from ever more credible witnesses, it seems to be yet another one of those subjects that mainstream science won't touch. Evidence for extraterrestrial visitation to the Earth in the form of UFOs is simply not looked at by science because, according to the 'best available scientific knowledge and expertise', it is highly unlikely that we are being visited.

It is the same kind of thinking which forbids the NASA science teams from looking for archaeological evidence on Mars: it can't be possible, therefore they won't look for it—even when the evidence for that possibility, as with UFOs, is staring them right in the face, as we have seen with many of the images presented in this book.

The nature of the problem is expounded upon by Professor Frank B Salisbury, Ph.D., Utah State University who did pursue some research into UFOs:[70]

> I must admit that any favorable mention of the flying saucers by a scientist amounts to extreme heresy and places the one making the statement in danger of excommunication by the scientific theocracy. . .

But one thing is for certain—*someone* is looking for, finding and assessing this evidence, and they are keeping quiet about it. Perhaps they are peering over the shoulders of NASA, or have their own means of studying these matters through other technologies and accessibilities. And perhaps too, the pot is beginning to boil because of a highly interesting development regarding one of the moons orbiting Mars...*Phobos*.

Phobos: Martian Moon or Artificial Satellite?

29. Phobos the largest Martian moon: only 15 miles wide
Image credit: NASA/JPL/University of Arizona

Mars has two moons, called Phobos and Deimos. They are quite unique in the solar system being the tiniest satellites of any planet, especially Phobos, which also orbits a mere 4,000 miles or so above the planet, far closer than any other moon to its host planet. Some theorize that, because they are so small, they must be captured asteroids, although apparently, the math doesn't add up for this to be viable. One scientist does have another theory, though.

Astrophysicist Dr. Iosif Samuilovich Shklovsky, who co-wrote the famous book *Intelligent Life in the Universe* with the late Carl Sagan, had been wondering why the orbital velocity and position of the Martian moons did not tally up with their mathematically forecasted positions, and after studying the gravitational, tidal and magnetic forces involved he concluded that there was no natural explanation for either the origins of the two moons or their extremely odd orbital behavior.[71]

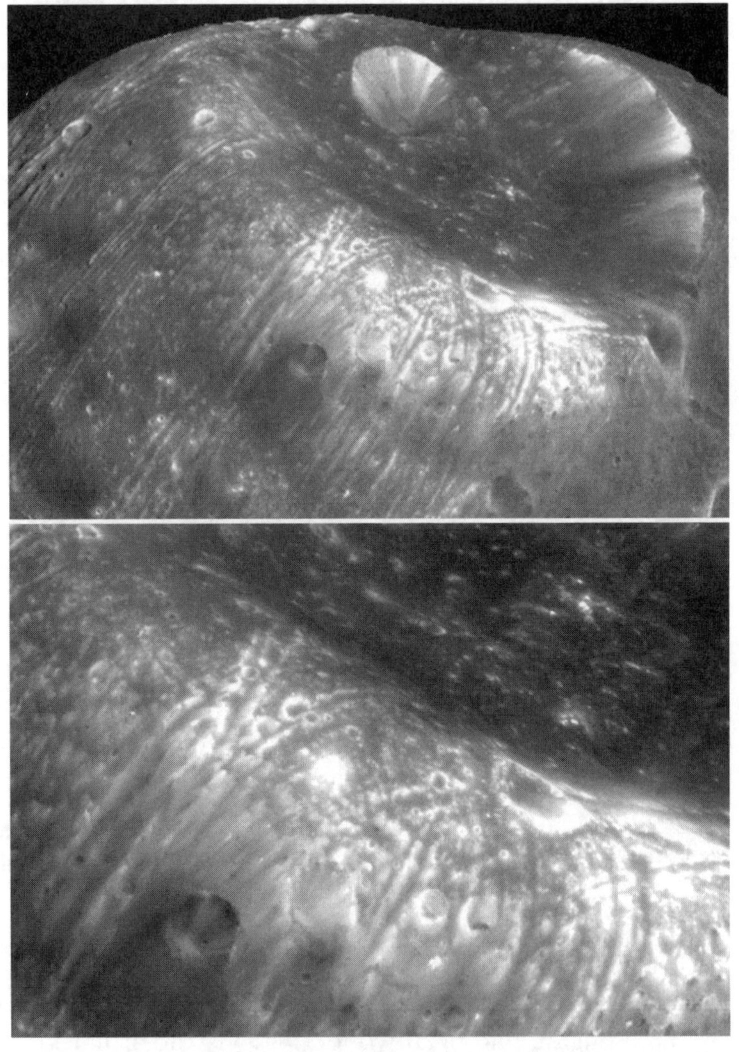

SM86. Phobos: structure revealed beneath surface dust? — HiRISE/MRO 2008

Image credit: NASA/JPL/University of Arizona

And he was particularly troubled by Phobos because, after he calculated the moon's orbital motion and studied its inexplicable accelerations, he came to the extraordinary conclusion that Phobos must be *hollow and empty* much like a tin can and could well be an artificial satellite orbiting Mars.

Even the United States Naval Observatory stated that Phobos is likely to be hollow if the accelerations were accurate, as was also concluded by the radio science data from the Phobos flyby mission, which stated that Phobos was unlikely to be a captured asteroid and that its interior is likely to contain large voids.[72]

So, we have an object with a highly unusual orbit around Mars, that is unlikely to be a captured asteroid and which has about a one-third hollow interior. Is there anything else that is unusual about Phobos? What about its surface? Can we see anything that might support Dr. Shklovsky's view that it must be artificial?

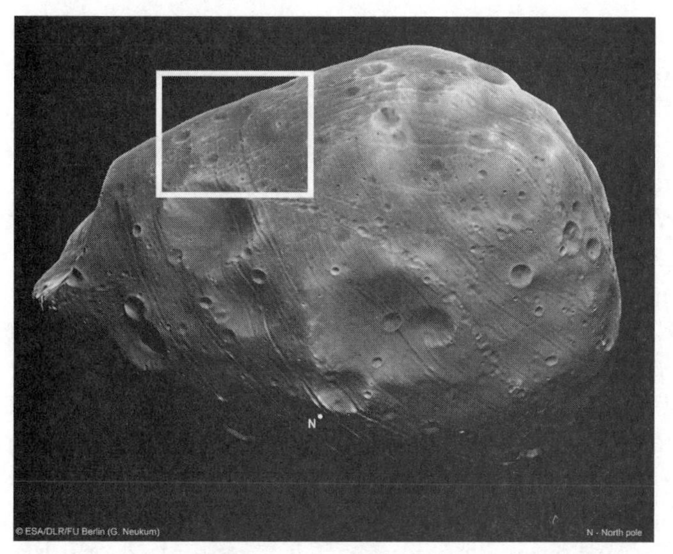

SM87. Phobos: rectilinear grid pattern — Mars Express
Source image credit: ESA/DLR/FU Berlin (G Neukum)
Graphic: SecretMars

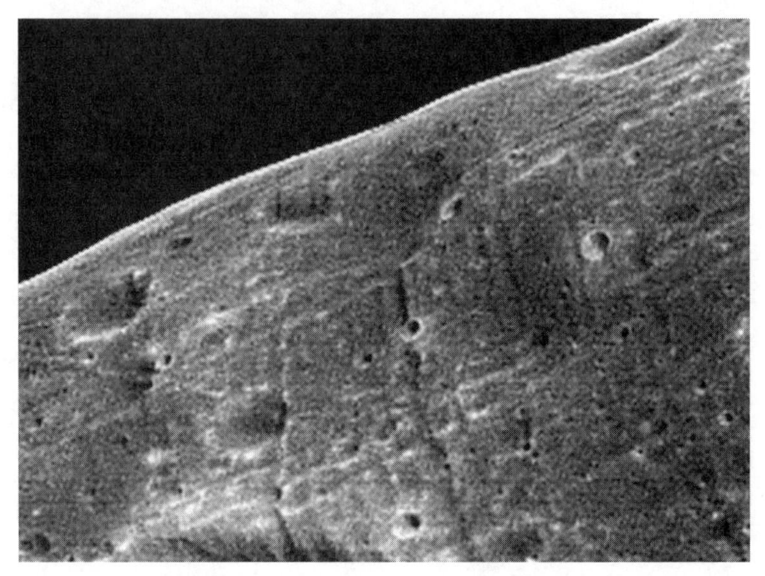

SM87.1 Phobos: rectilinear grid pattern (detail)
Found by Richard C. Hoagland

First highlighted by Hoagland,[73] there is a portion of this Phobos image that we found to be very interesting. And yes, the old 'geometry' favorite comes up again as we see familiar repetitive, grid-like patterns stretching across a large area of the moon.

What are such markings doing on a hollow, 15-mile wide rock orbiting the planet Mars, we have to ask. Could they simply be following an underlying structure buried beneath and thus reveal the unmistakable signature of intelligent construction?

If Dr. Shklovsky is correct and this is an artificial body—an orbiting satellite or spaceship—it looks to be in extremely old condition; battered, pock-marked, cratered, and covered by many layers of Martian and cosmic dust, built up over thousands or even millions of years. Either that or it is a hollowed-out rock engineered for its purpose. Whichever the case, we would presume it to be long abandoned.

Is there anything else about Phobos which might support the artificial construct hypothesis?

Well, there is *the monolith*…

In 1998, independent researcher Efrain Palermo[74] found an unusual object on the surface of this moon that raises a few questions, its uniqueness confirmed by Lan Fleming an imaging sub-contractor at the NASA Johnson Space Center.

Named "The Phobos Monolith" after the famous alien object featured in Stanley Kubrick's film, *2001: A Space Odyssey*, it stands some 76 meters/250 feet high, eerily and completely isolated, with no similar-sized objects anywhere to be seen.

With its long shadow and distinctly geometric shape, it strikes a deeply enigmatic profile.

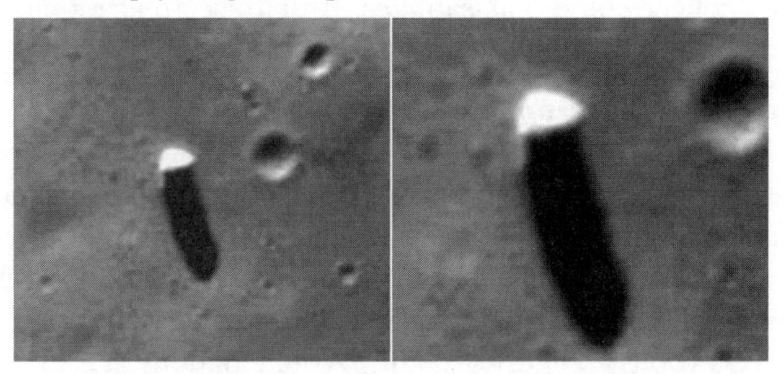

SM88. The Phobos monolith — MOC/MGS Orbiter
Region: Phobos Moon, Sep 1998; found by Efrain Palermo
Image credit: NASA/JPL/MSSS

In Kubrick's film, inspired by Arthur C. Clarke's short story "The Sentinel", an ancient alien race had visited our solar system at the dawn of humanity and had purposefully left a mysterious black monolith on our own Moon orbiting the Earth. The purpose being that, when humanity finally made the leap into space, reached the Moon, and discovered the monolith, the alien race would be alerted to this event, the monolith acting like a 'cosmic alarm bell'.

What do creative writers tap into now and then regarding the deeper unseen mysteries of life? Can they know something beforehand in a prescient, intuitive way, I wonder?

No surprise then that Phobos has drawn the attention of the world's space agencies. A hollow moon could provide the perfect base for exploring Mars. But it could also harbor the most significant find ever made in the history of the human race for what if, within its hollow chambers, a treasure hold of advanced extraterrestrial technology was to be found? That would be an enormous prize worth something to some nation and its corporations.

Are there any spacecraft planning to go there? Well, the Russians sent *Phobos-Grunt* in November 2011, but it became stranded in Earth orbit for over two months before it finally crashed into the sea. Its official mission was to land on Phobos, collect some soil samples, and then return them to Earth.

Do Space Missions Get Sabotaged?

After the Russian mission to Phobos had failed, conspiracy theories were raised in that the mission might have been sabotaged by the Americans. Well, it may be unfair to level such an accusation, but it might be worth pondering what the stakes could be now if Phobos is exactly what the Russian astrophysicist thinks it is. Would nations start taking potshots at each other in the rush to be the first to get hold of alien technology? Wars have been started over much less.

The Russians have had miserable luck concerning their missions to Mars, and if the Americans should have their political sights set on being the ones to dominate exploration of the Red Planet it must be considered plausible that they would want to discourage and perhaps even derail missions from other nations that might interfere with their goals and ambitions. They certainly don't shy away from using subversive means to assert their worldwide dominance in

Earthly affairs, so I see no reason why they would decide to alter their behavior concerning the new frontier of space.

I must admit therefore to not having been surprised when the European mission *Beagle 2*, crash-landed on Mars during December 2003 in the region of the planet, known as Isidis Planitia. It was an *astrobiology* mission, designed to detect life.

NASA, as we know, has preferred to avoid life detection missions, and I recall being curious to see how the Americans would react if the British-led science team on the Beagle mission were to land on Mars and beat them to the history books by discovering life—especially after all the dollars NASA had invested in their decades-long space program. As it turned out, they did not have to worry about it.

Ancient Intelligent Life in the Solar System

The problem we had with Mars was in conceiving that an intelligent civilization could have been living there way before any signs of intelligent life ever appeared on the Earth. The argument being, that any potential ruins identified by NASA images must have been constructed at a time when Mars had breathable air—something which may not have been possible for millions of years on Mars, making the ruins therefore extremely ancient.

But that is not the only scenario, because, as we have already suggested, there is also the possibility that colonists may have arrived when Mars had no breathable air and simply constructed their buildings either on the surface or perhaps underground, as soon as they arrived. Which means that any Martian ruins found may not necessarily have to be millions of years old, but instead much more recent. Even so, the idea that an intelligent civilization was in existence in our solar system way before our human one, is challenging.

Could intelligent life have evolved on other planets orbiting our Sun, before it arrived here on ours? Well, if Earth is considered to be the perfect planet in our solar system for intelligent life to have evolved, due to its ideal distance from the Sun—being not too hot or too cold (what's referred to as the "Goldilocks Zone")—then logically the answer must be no, and we should disregard the other planets as candidates, as they are orbiting in less habitable zones.

Which again brings us to the possibility that, if evidence of intelligent life were to be found elsewhere in our solar system, such as on Mars, then it would be much more likely that this intelligence would have originated from *some other solar system* rather than having indigenously evolved upon planets within our own.

And if the idea of ancient extraterrestrial visitors to our family of planets still strikes skepticism within your mind you may wish to consider the following object, which might just blow any doubts you have as to this possibility completely out of the water. It is the most extraordinary anomaly and in times to come may provoke the most profound questions yet to be asked concerning the origins of intelligent life in our solar system…

The Mystery Moon of Iapetus

This is the enigmatic moon *Iapetus*, which orbits the ringed planet, Saturn. It possesses the most striking linear feature yet seen in the entire solar system, being encircled by a *hemispherical wall* about 11 miles high and wide. It appears there is nothing in geology yet that can explain such a clearly-defined and geometrically spectacular oddity, which suggests that the moon Iapetus could be an artificial construction—and a huge one too.

The question as to who could build such a thing, and for what purpose, would open a Pandora's Box of wild speculation. The similarity to the "Death Star" from the Star Wars movies is both provocative and just a little disturbing. We might presume that it must be very ancient. We would just have to land on it, find the door and take a look inside. We should also expect that such a mission is already being planned.

30. Iapetus: natural or artificial moon? — Cassini Orbiter
Image credit: NASA/JPL/Space Science Institute

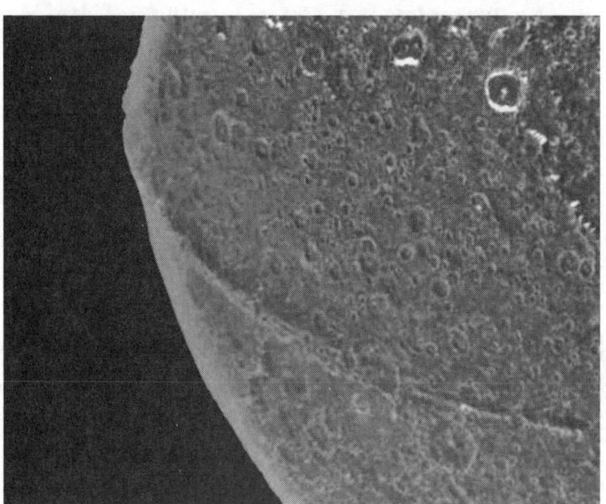

SM89. The Iapetus wall — Cassini Orbiter 2004
Image credit: NASA/JPL/Space Science Institute

And if this completely unnatural-looking gigantic wall is not enough to start wondering about moon-sized spaceships, observe what Hoagland noticed about the *shape* of this object (SM90). This image of Iapetus obtained by using the glare of Saturn, dramatically reveals that it is not even a round sphere, like most other planets and moons, but instead has incredibly straight edges as seen here on the horizon—which can only indicate that this moon's surface is *multifaceted!*

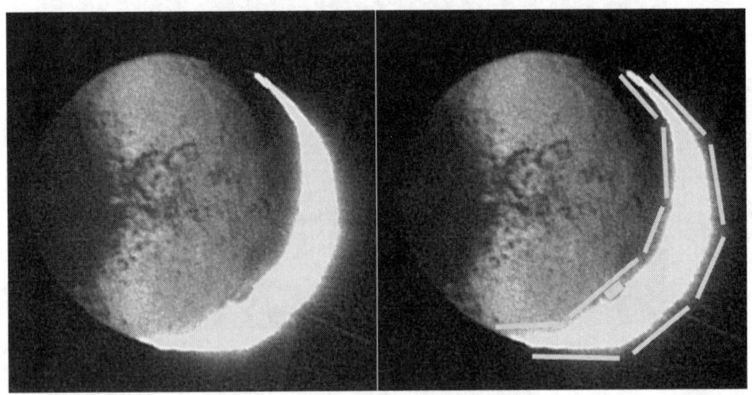

SM90. The geometric horizon of Iapetus — Cassini Orbiter
Found by Richard C. Hoagland
Source image credit: NASA/JPL/Space Science Institute
Graphic: SecretMars (from Hoagland's original concept)

There can surely be no clearer evidence that what we are looking at is an artificial object, an 'orbiting space station', or something of that nature. And as with the Phobos orbiting anomaly, these appear to be very, very old at least from what we can observe from their outward appearance. There should remain however the possibility that these are not derelict relics from a long-forgotten time at all but may, in fact, be operational. Is NASA aware of this data concerning Iapetus? Of course, they are. And those who govern us will have been informed about it.

So now that we have two pieces of pivotal evidence suggesting that there are very advanced and possibly very ancient technologies orbiting around the planets Mars and Saturn, what else might be lurking within our solar system?

How about we move a bit closer to home…

Our Secret Moon

31. Astronaut Buzz Aldrin: stepping onto the surface of the Moon from Apollo 11, 1969
Image credit: NASA/JPL

Much has been much written about conspiracies concerning the Moon ever since Apollo 11 landed there in July 1969 and Neil Armstrong set down the first human foot on its surface. Armstrong himself uttered a very cryptic remark during a speech given in 1994 at a special White House celebration of the 25th anniversary of the Apollo 11 mission; a thinly-veiled comment possibly suggesting that he and the other astronauts were not allowed to speak freely about their experiences and what they had really discovered while walking on the moon, he said:[75]

There are great ideas undiscovered, breakthroughs available to **those who can remove one of truth's**

protective layers. There are places to go beyond belief because there lays human destiny. [Emphasis added]

Why would he say such a thing as removing "Truth's protective layers" when talking about the Moon? Could the Apollo astronauts have witnessed something extraordinary which they were forbidden to discuss?

Hoagland, in his ground-breaking video presentation *The Moon-Mars Connection*,[76] revealed a completely alternative view of our Moon that goes far beyond the mere crater-battered and unremarkable cosmic body that NASA presents to us. His digital enhancements of photos taken by the lunar orbiting spacecraft and by the astronaut's cameras on the surface revealed detail in the lunar sky possibly suggesting the existence of huge eroded structures several miles high—remnant edifices that would have to be very ancient indeed.

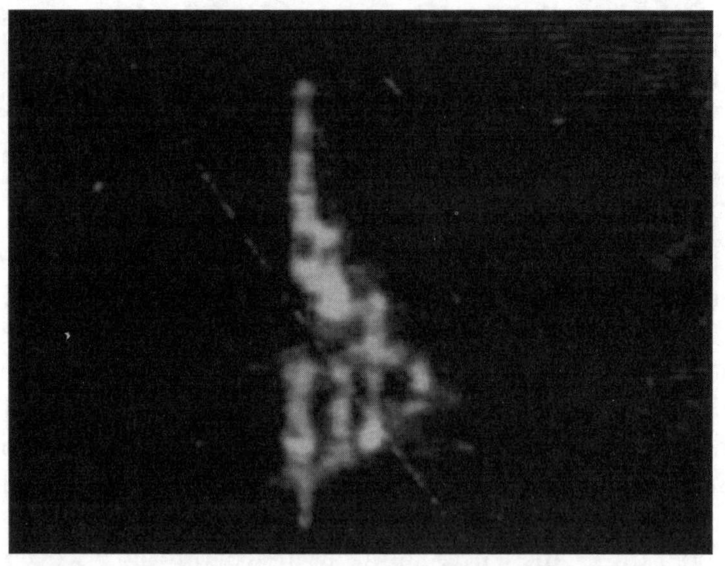

SM91. The "Castle" discovered by Richard Hoagland in a photo from Apollo 10. This image has 'disappeared' from the NASA NSSDC database and can no longer be found
Image credit: NASA/JPL?

His theory is that there are several massive domes, on the Moon, possibly made of glass—which in the Moon's environment has incredible strength, apparently—and which over many thousands of years of cosmic bombardment from meteorites have been reduced to a shredded, thin veil of crystalline mesh that, depending on the angle of sunlight, will occasionally reveal its presence in some of NASA's images, particularly those taken of the horizon.

Hoagland found what he believed to be a remnant piece of one such dome in the form of a remarkable object suspended nine miles above the lunar surface, which he named the "Castle" (SM91). The interesting thing about this object apart from its spectacular geometric structure is that he says he found the image on a photographic print of the original NASA frame AS10-32-4822, but when further copies of the same frame were ordered, the object had vanished.[77]

It transpired that there were several frames associated with the number 'AS10-32-4822', probably due to its origin as a power winder sequence that included several pictures taken, one after the other. This might explain the Castle object's disappearance due to the angle changing from one picture to the next and the object no longer reflecting sunlight. It may also have just been a scratch or mark on the film negative, and therefore didn't exist. However, the crucial factor here is that neither possibility could be verified because the original picture from which Hoagland's print showing the Castle came had itself *disappeared from the National Space Science Data Center (NSSDC) database.*

Which offers another explanation for The Castle's vanishing act . . . someone at NASA is very sensitive to unusual artifacts appearing on their photos and will go to some lengths to remove them from the public eye by airbrushing them out or even removing the original film negatives from the official archives!

NASA Falsifies Official Moon Images

Surely a wild accusation you might think. NASA would never airbrush things out they don't want us to see and then restore the falsified data into the image library for the public and scientists to study. But Ken Johnston, who said he was the supervisor of the data and photo control department in NASA's Lunar Receiving Laboratory during the Apollo program, states this very thing has happened.[78]

Ken Johnston is a retired aerospace engineer and a former U.S. Marine and civilian astronaut consultant pilot from the Apollo Moon Program. While working for a NASA contractor in the Lunar Receiving Laboratory, he claimed it was his responsibility to catalog and archive all of the many thousands of Apollo orbital and handheld photographs as well as the 16mm films from the command and Lunar modules which often filmed sequences from the lunar orbits, descents, and ascents. One of his duties was to arrange the screening of these films for science and engineering teams and on one such occasion, he said something extraordinary happened.

During the playback of a film from the Apollo 14 mission he saw several lights inside the rim of a crater, and a strange "Column or plume—or out-gassing or something", rising above the rim. One of the scientists viewing the film—chief astronomer Dr. Thornton Page—had him stop the film, rewind it and repeat the sequence of the 'plume' several times.

The film was later returned to the NASA photo lab for storage and was to be checked out again the next day for it to be shown to some lower-ranked scientists and engineers. While replaying the film for his new audience Johnston was shocked to discover that the lights and out-gassing had disappeared from the film! Later that day, he came upon Dr. Page and asked him what had happened to that unusual sequence, receiving a grin and the reply:

"There were no lights. There is nothing there."

Obviously, overnight at the NASA photo lab, someone had been told to 'clean up the film' and remove the mysterious goings-on at the crater—which has enormous implications. For suddenly we have an ominous dark cloud looming over the honesty and integrity of the entire NASA and American space program. I mean, what did NASA find on the Moon that was so important that they would sanction falsifying the official science data photographs and films of the Apollo 14 mission and deny science and the public the truth about what they discovered there?

This situation would imply that:

- NASA knows a great deal more about what's on the Moon than they have declared to the world and that this knowledge is classified above top secret.

- NASA astronauts were under strict orders not to discuss certain things they saw.

I have heard stories that I use to simply regard as myths, such as when Apollo 11 landed on the Moon there were huge spaceships lined up near a crater watching the proceedings. I dismissed the story, but now I start to wonder. It is said that NASA used to have a code word which astronauts were to use on audio transmissions whenever they saw a UFO in space . . . "Santa Claus". No doubt the secret word has been changed to something else by now.

A Secret Base on the Moon

So, what could NASA be hiding on the Moon, something they don't want you and I to know about?

In 1965, Sgt. Karl Wolfe, a precision electronics photographic repairman with top-secret crypto-clearance, was loaned to the Lunar Orbiter project to repair some equipment that was malfunctioning. An airman who was explaining to Wolfe how the equipment worked paused to say in a distressed manner something extraordinary:[79]

"By the way, we've discovered a base on the backside of the moon."

At which point the man proceeded to lay out some photos which clearly showed, "Structures, mushroom-shaped buildings, spherical buildings, and towers." Sergeant Wolfe recalled feeling frightened at that time because he knew they were working in compartmentalized security and that the airman had breached that security by showing him those photos, and so he did not question him again. He worked there for three days and afterward went home thinking:

"I can't wait to hear about this on the evening news!"

Karl Wolfe never did get to hear about it on the news, and neither did anyone else, until 2001 when he presented this information to the Disclosure Project National Press Club Conference in Washington D.C. There he declared before the world's media, that he would testify before the U.S. Congress that what he said was the truth.

A *base* on the far side of the *Moon*? How could the human race have possessed the technology to build a base there in 1965, when they had not even sent any astronauts yet to even land on it? Apollo 11, with Armstrong and Aldrin, didn't touch down on the surface of the Moon until some four years later. Sergeant Wolfe's testimony, therefore, suggests **four** possibilities as to the reality of such a Moon base:

1. **There is no base**—Sergeant Wolfe had been set up and the photos of structures and buildings were fakes. It was a deliberate ploy to plant a disinformation myth that the Moon was inhabited by aliens. At the time, the USA was in the midst of the Cold War era with the Soviets and official documentation has suggested that UFO sightings were being used by the CIA as cover stories for top-secret military projects.

2. **The base is real and is being used by the Americans**—advanced classified technology has enabled them to travel to the Moon and build a base there

in secret. The purpose thereof to set up a military facility or a scientific station to possibly study ancient ruins that have been discovered, out of sight from the public space program that uses inferior technology.

3. **The base is real and is derelict and long-abandoned**—an ancient civilization built it that has since disappeared.

4. **The base is real and is being used by an extraterrestrial civilization**—they are probably using it as a base of operations for a long-term mission to study Earth and humanity.

Given the evidence so far, what do you think is more likely?

Ken Johnston and Karl Wolfe were not the only ones to witness NASA's censoring of official photos and scientific data. Donna Hare was employed by NASA for 15 years as a sub-contractor, a photo technician during the Apollo mission years. She had a secret clearance which enabled her to go to a restricted area where they developed pictures shot from satellites and the Apollo missions.

One day she was talking with one of the photographers and developers there when he drew her attention to a photo mosaic he was working on. There was a white oval-shaped object that was casting a shadow on the ground beneath it. She asked him if it was a UFO, to which he replied, "Well, I can't tell you." So she asked what was going to be done with this information, to which he answered:[80]

"Well, we have to airbrush these things out before we sell these photographs to the public. . ."

So, what could the Apollo astronauts have possibly discovered on the Moon?

Artificial Objects on the Moon

Apollo 17 was the last manned mission in 1972, during which astronauts Eugene Cernan and Harrison Schmitt landed at the south-eastern edge of Mare Serenitatis to explore the Taurus-Littrow Valley. On the second of their Moon Buggy excursions, they investigated a crater the astronauts named "Shorty". When describing the region they were exploring as he drove the buggy while Schmitt was taking photos, Commander Cernan had earlier that day exclaimed:[81] "Man, you talk about a mysterious looking place. . ."

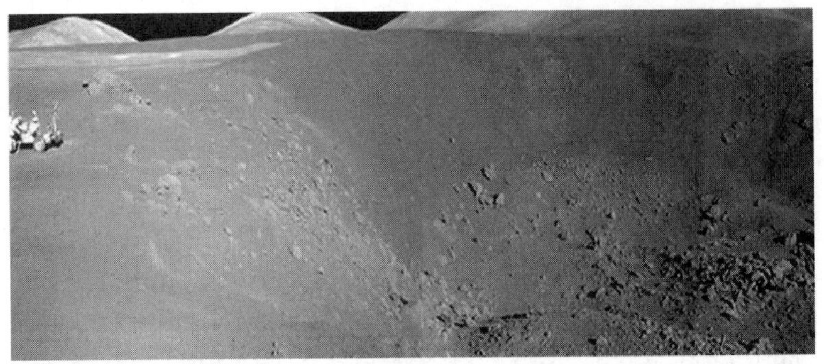

32. Lunar Rover Vehicle at Shorty Crater — Apollo 17, 1972
Source image credit: NASA/JPL; composite picture: SecretMars

Arriving at the 110 meters wide crater, the astronauts looked into it and appeared to be impressed by what they saw. Cernan saying to Houston:[82]

"This [crater] is an impressive one. Wait until you see the bottom of it..."

Photos of the crater were taken and what we can see now in the close-up of the crater floor is remarkable (SM92).

The detail is not great here, and someone with more image enhancement skills than I possess will get better results, but I think there is sufficient consistency of objects with geometric and linear characteristics to denote that we are probably looking at a shattered debris pile, perhaps both of masonry and technology. I say masonry because I get the feeling of perhaps remnant ornate architecture is evident here.

This is also the location where Hoagland found what he described as a possible 'robot's head', comparing it to the likeness of the Star Wars movie character, C-3PO!

SM92. Lunar junkyard at Shorty Crater? (110m)
Region: The Moon, Dec 1972; image credit: NASA/JPL
Graphics: SecretMars

Based on the simple contrast enhancement I used on this picture (SM92.1) I do not see any clearly defined features to merit labeling it as an artifact. However, Mike Bara and Richard Hoagland as presented in their book *Dark Mission* used more advanced image enhancement techniques that appeared to extract more interesting detail, strengthening the case for the possible artificial nature of this object.[83] The fact that this thing lays within a site on the Moon that includes several other artificial-looking objects would, of course, increase the odds for this too belonging to that category.

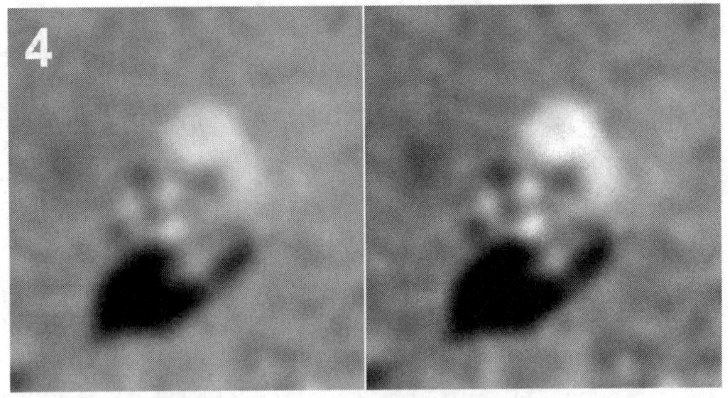

SM92.1 Object 4: detail of "Robot head" (15-20 cm)
Found by Richard C. Hoagland

But did the astronauts go down there in the crater and take a closer look? Well, according to the audio recording of their activities, they didn't.[84] In fact, given the steepness of the 14-meter deep crater and their mission time constraints, it was perhaps unlikely they made their way down the crater slopes and therefore did not see the objects revealed here in the photo close-ups. Also, while they were photographing the crater, geologist Harrison Schmitt discovered some orange-colored soil which seemed to get them both excited, and according to the audio transcripts this discovery then took up the rest of their time and attention before they left Shorty.

The next day, however, on their third EVA, they stopped at a large boulder near *Henry Crater* and took some photos of the area. Several images captured a very unusual object on a distant slope, which appears to be a very mechanical or architectural-looking anomaly (SM93), although the astronauts made no comment about it.

The symmetrically placed 'holes' on the left-facing side of the object together with its curvature and linear elements, make it certainly very artificial-looking (not unlike an engine piece—and well over a meter in size). If only the astronauts had gone over to take a look.

SM93. Artificial object at Henry Crater? (2m) — Apollo 17
Found by Richard C. Hoagland
Source image credit: NASA/JPL; graphic: SecretMars

It is absolutely fascinating to imagine what ancient event may have brought this possible artificial object to its final resting place here in the lunar dust; perhaps a ghostly artifact from some unknown civilization that once graced our Moon.

But if we find one potential artificial object at a specific location it behooves one to have a look around for others in the vicinity, proceeding on the premise, as we did on Mars, that 'something' may have happened at this site a long time ago that resulted in scattered debris. It strengthens the case considerably to find corroborative evidence nearby, and upon examining photos of the sloping hills close by a couple of unusual shaped, smaller objects that could be related to the larger discovery, did catch my eye (SM94).

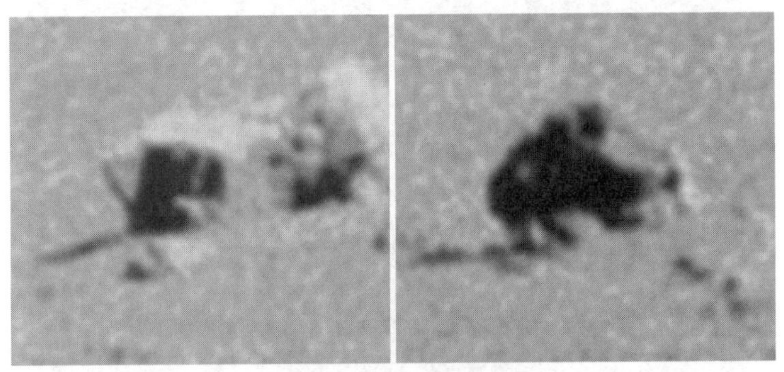

SM94. More artifacts near Henry Crater? (15-20cm)
Image credits: NASA/JPL

Did NASA astronauts go to the Moon and discover evidence of a previous civilization there? And has NASA been sitting on this knowledge ever since?

The evidence is strongly suggesting this. And with NASA photo labs in the loop too censoring and doctoring pictures for public and mainstream scientific consumption, it is plainly obvious that the American space agency is deeply involved in a very serious and secret agenda.

The Extraterrestrial Reality: Time for Disclosure

So now the pressure appears to be mounting for the chain of command to come out and release some real information on this subject. Constant public denial and humorous dismissive quips from elected politicians on the subject of 'aliens and little green men' no longer washes. People are not stupid and they are beginning to see through the non-sense. The internet has also enabled information to spread like wildfire, leading the public to become even more in-formed than those who represent them in government, which has led to a massive swell of impatience with the po-litical response received on this subject.

This lack of official acknowledgment of an intelligent ex-traterrestrial presence in our solar system first led to "The Disclosure Project", led by Dr. Steven Greer, giving a worldwide presentation to the world's media at the Wash-ington Press Club in 2001,[85] where twenty high-caliber wit-nesses presented testimonies describing their experiences of contact with aliens, UFOs, and secret technology.

However, not producing the desired response from the United States government, another people-led movement had another go in April 2013, when, during a five-day hear-ing, some forty researchers and military/agency witnesses testified before former members of the U.S. Congress. Called "The Citizen Hearing on Disclosure of an Extrater-restrial Presence Engaging the Human Race"[86] the attempt was made to present the facts concerning extraterrestrial contact and government knowledge of such events.

The international spokesperson for this hearing, which was again held at the Washington Press Club, was Dr. Ed-gar Mitchell (Capt. USAF, NASA Apollo 14 astronaut). Dr. Mitchell had been prominent for several years in pushing for the truth regarding the extraterrestrial presence to be put before the American people, and has said:[87]

> We're way, way past time for bringing this information to the public, acknowledging it and entering into the sort of discussions at the highest levels of Congress and the government — to make this information available to the public. It's far, far past time for this. If we don't do it

here, I don't know when we'll ever going to be able to do it.

It certainly is time. When you start to put things together, such as Ken Johnston's testimony of the deliberate faking of evidence of intelligent activity on the Moon, the disappearance of the original NASA photo that showed Richard Hoagland's 'Castle', Sgt. Wolfe's secret moon base photos, Donna Hare's experience of photo censorship of a UFO, and of course the avoidance by NASA to discuss all the strange objects and structures found on the planet Mars, a consistent picture emerges; one of obfuscation, denial, and a deliberate, consistent policy of secrecy.

It all begins to tie together too when we start to listen to what NASA astronauts have been saying over the years. Perhaps now they have been given permission to talk about this stuff; a whisper in their ear that they won't be subject to prison and penalties anymore. Well, maybe in some shady corridor it has now been decided that some information needs to be revealed to the people about these matters, gradually, carefully and in drip-feed fashion, so they can perhaps slowly get used to the idea that 'alien civilizations' inhabit this universe with us . . . and have been visiting the Earth for some time.

Astronauts and UFOs

At the forefront of space-discovery have been NASA astronauts and Soviet and Russian cosmonauts. It has been these guys who have been beyond the sphere of our home planet and ventured out into the mysterious, forbidding world of space. If there are deep mysteries to behold beyond the confines of Earth then it will be to the words of these bold people that we might listen to with all seriousness if we want to learn something.

All of the early Apollo astronauts were military personnel sworn to oaths of secrecy, but when they approach their later years in life something within them starts to loosen up it seems, or they appear perhaps less fearful and more open to sharing their experiences of what they saw and witnessed in space. Either that or, as we said before, they have now been permitted to talk about this stuff.

Let's hear what our pioneering spacemen have to say about UFOs and the reality of extraterrestrial intelligence visiting our planet:

Astronaut Eugene Cernan, who commanded Apollo 17, made this comment about UFOs:[88]

I've been asked and I've said publicly I thought they were somebody else, some other civilization.

Astronaut Dr. Edgar Mitchell, from Apollo 14 and sixth man to walk on the Moon declared:[89]

We all know UFOs are real. All we need to ask is where are they from . . . the evidence points to the fact that Roswell was a real incident and that indeed an alien craft did crash and that material was recovered from that crash site.

Former NASA astronaut and Princeton physics professor, Dr. Brian O'Leary:[90]

There is abundant evidence that we are being contacted. That civilizations have been visiting us for a very long time . . . these visitors use the technologies of consciousness.

Gemini astronaut, Colonel Gordon Cooper in a letter to the United Nations:[91]

I believe that these extraterrestrial vehicles and their crews are visiting this planet from other planets which obviously are a little more technically advanced than we are here on Earth. I feel that we need to have a top-level, coordinated program to scientifically collect and analyze data from all over Earth concerning any type of

> encounter and to determine how best to interface with these visitors in a friendly fashion.

He also stated:[92]

> We know the government is sitting on hard evidence of UFOs.

Astronaut Storey Musgrove, who flew on the space shuttle repair mission of the Hubble Space Telescope:[93]

> Statistically, it's a certainty there are hugely advanced civilizations, intelligences, life forms out there. I believe they're so advanced they're even doing interstellar travel. I believe it's possible they even came here.

Do we get the impression that something monumental is going on in space but news of which is just not reaching people in the world? How can this be? Is there a news blackout? These are the views of highly trained astronauts and scientists with impeccable credentials. They speak of a reality that sounds like science fiction, but which must in truth simply be relating the sober reality of life out in space.

With Earth's astronauts acknowledging the possibility and even the probability that extraterrestrial spacecraft are visiting the Earth, the likelihood of a base on the Moon, therefore, becomes a serious possibility. Why? Because on a purely practical level you would expect that any visitors from a far-distant star system would require a *local operational base* from which to fly their reconnaissance missions—and the Moon would provide an ideal observation station to monitor the Earth.

Over the years there have been many sightings of luminous objects shooting across the surface of the Moon under seemingly intelligent control, indicating there is a continuous activity in the lunar skies. Numerous NASA images show too there is plenty of evidence for artificial structures on its surface, reminding us of the kind of thing we've already seen on Mars. This suggests we may well be just as likely to find inhabited buildings on the Moon, as well as ancient ruins.

Do Covert Agencies Control Alien and UFO Affairs?

What is to be done with this state of affairs? On the one hand, we have this massive body of evidence from both highly credible witnesses and official photographic and digital images affirming the highly probable existence of extraterrestrial craft, buildings, and artifacts—and on the other hand, a completely unmoved and unresponsive government and space agency, denying that anything of any serious note or scientific interest is happening or has been discovered at all. It is incredulous.

The reality must be that someone somewhere is struggling to know what to do with this information, and perhaps everyone is avoiding dealing with it, for the consequences of releasing knowledge of such matters to the world will be enormous and paradigm-shattering. There is strong evidence too that, due to severe compartmentalization within hierarchal structures that control the flow of information, knowledge, and control of alien matters may have been gradually siphoned off to covert, clandestine agencies that may not be under the direction of legitimate government but which serve some other goal.

Dr. Steven Greer from the Disclosure Project has compiled evidence that links prominent U.S. hi-tech companies and defense contractors such as Bechtel, Boeing/McDonnell Douglas, Lockheed Martin, SAIC, BDM, Northrop Grumman and Wackenhut with the development of technologies derived from the back-engineering of crashed alien spacecraft recovered by the military.[94]

Former U.S. Army Sergeant, Clifford Stone confirmed the existence of such military units specifically tasked to secure these downed extraterrestrial spacecraft:[95]

> Under [Project] "Moon Dust" and under "Blue Fly", we have recovered alien debris, not of this Earth . . . I was involved in those types of operations to retrieve crashed E.T. objects . . . You would go ahead and deploy as though it was a nuclear accident. There are procedures already established on nuclear or biological or chemical accidents . . . those same procedures are utilized when you do a recovery or extraction of a crashed E.T. spacecraft.

With such an extraordinary situation in operation where military projects exist specifically to recover extraterrestrial craft, this suggests that if the U.S. government is not in the loop regarding the existence of such craft and what is being done with them, then the door has certainly been left open for *someone* to exploit these alien vehicles and their priceless technology. If the military is in control of recovering the crashed (or targeted) UFOs, are the private aerospace corporations handling the research, development, and commercial exploitation of them?

Yes, 'commercial exploitation'. The business of grabbing alien spacecraft for profit would be lucrative beyond imagination. For if we are talking about the superior, technologies used by extraterrestrial civilizations that enable them to traverse between solar systems, zip around our planet at speeds impossible for conventional aircraft, as well as to cloak and disappear at will as has often been reported, then their technology is going to be worth a hell of a lot to those Earth companies seeking to dominate their markets.

However, it goes without saying that the acquisition of extraterrestrial vehicles and their advanced capabilities, would be of paramount interest first and foremost to the military. But who would, in turn, still require the skills and expertise of the corporate defense contractors—those who normally build their weaponry and aircraft—to develop any such newly-discovered alien technologies into useful applications.

An insight into such an industrial arrangement is provided by **Colonel Phillip J. Corso, Sr., U.S. Army (ret.)**, an Army intelligence officer who was on the staff of President Eisenhower's National Security Council. He also worked in research and development and revealed that advanced technological information recovered from extraterrestrial spacecraft *was deliberately fed to commercial corporations:* [96]

> We gave information about the technologies out and insisted they [the corporations] take the patents. But also we put a little bit of a requirement: Feed it back to us, the competitive edge of the Army—take the patents, make all the money you want—but give it to the American people and give it to the world.

So it would seem from this statement that the companies took the technology, developed it, gave some advanced capability back to the military for their projects, and then turned it into a product for sale to the world. It seems like a win-win situation.

But around this time, President Eisenhower was becoming deeply concerned by his lack of control of this 'military-industrial complex' that was growing from the exploitation of alien technology, as we previously saw in his final address to the nation. What was it that worried him?

Did the corporations start holding back some of their discoveries derived from the E.T. technology, perhaps not sharing with their military gift-horses all of their breakthroughs? Did they no longer wish to wait for piecemeal handouts, but instead became proactive in seeking E.T. technology for their own corporation's explicit benefit and control? Taking whatever steps were necessary, in their eyes, to jealously guard and secure their extraterrestrial prizes?

Over fifty years have passed since Eisenhower's warning to America and who knows what has transpired since then regarding this situation. Today, however, there seems to be a quagmire of interwoven secrecy agendas that are in competition with each other, making it difficult to unravel and discern who exactly controls this highly classified sector that exists somewhere in the USAP world (Unacknowledged Special Access Program), deeply buried and untraceable.

There also seems to be a movement afoot by higher echelons of the military and intelligence community, suggesting they are in opposition to these covert programs and who want the secrecy ended. This suggests the military may indeed have somehow lost control of that technology and want to rectify the situation, although the public has yet to be informed about what they were even doing with E.T. technology way back then and how they got it.

A lot of catching up to do regarding how this history has unfolded over recent decades, but I would guess that 'planetary defense' was top of their list back in the 1940s-50s and if that security has somehow been put in jeopardy by groups that now have superior technology to the military . . . well, a complex and dangerous state of affairs may be imagined to be at the heart of all of this.

To be honest, I don't have complete answers yet to the dark, foreshadowing questions that embroil this portentous matter, but I hope in the near future to be able to throw some light upon it. However, to discover the truth, we depend on people coming forward to share with the world what they know, and until some kind of disclosure from government happens that is the only way forward. If you know something important, we need you.

For now, let's imagine for a moment just how these global business corporations might truly react if E.T. turned up tomorrow with some far-reaching technology. To offer the human race some new, miraculous devices that might replace all of their outdated products.

I think there will be **two** chief concerns that might worry our global business magnates:

1. **A visit from a benevolent E.T.** who, on observing our human health and planetary ecological problems caused by lack of food, dependency on prescription drugs, etc., and our use of fossil fuels for energy, might, for example, offer humanity *free energy devices, food replicators, and full health restoration modules* for every person and home on the planet, and at a stroke, wipe out starvation from our world as well as our dependency on expensive medicines, oil, gas and electric. Such a scenario would completely freak out all of the pharmaceutical and energy companies.

2. **A visit from an opportunistic E.T.** who, on observing how our limited planetary resources were controlled by a handful of global corporations, decided to make a trade deal with one or two of them *to supply advanced technology*—that would freak out all of their political and industrial competitors.

So, if you have a bunch of boys and girls who don't want to lose their profits from their oilfields and drugs, and who also don't want to risk others getting hold of E.T. technology that might put them out of business forever, what would they in all likelihood do?

Well, concerning the benevolent E.T. offer to the entire human race, some of our more enlightened corporate leaders may, of course, choose to humbly acquiesce and embrace the whole wonderful change that such a beneficial gift from E.T. would bring to the Earth and humanity, and help us herald in a new age of prosperous health and a clean, renewable and limitless energy supply.

But given the level of self-serving greed and arrogance displayed by the majority of humanity's global corporations, most would surely view such a benevolent act only as a massive assault on their corporate profits and control, and as such would more likely embrace the business-orientated

E.T. visitor out looking for trade. Then, alien world-changing technologies could be purchased or licensed perhaps for exclusive manufacturing and distribution rights on the Earth.

Failing a trade deal with E.T., they would probably take steps to block such interference and 'aggressively acquire' the very technology E.T. wants to give us for free, as fast as possible and by any means at their disposal, before their competitors do.

It makes you wonder what the Star Wars orbiting weaponry program was all about back in the 80s. How about targetting and bringing down spacecraft, from friendly visiting peoples from other worlds, to recover and steal their technology?

Would make a cracking movie Mr. Spielberg…

Is it really possible that secret groups in league with corporations and the military are operating outside of government control and supervision and pursuing their own agendas? If so, it behooves our true world leaders to expose and reveal what they are up to, and what is really happening on this Earth.

Apollo 14 astronaut Edgar Mitchell sums up the seriousness of the situation:[97]

> Whatever activity is going on to the extent that it is a clandestine group, a quasi-government group, a quasi-private group, it is without any type of, as far as I can tell, of high-level government oversight. And that is a great concern.

And now, with this perspective, we might understand how and why the search for and discovery of possible alien civilizations on Mars and elsewhere in our solar system—indeed *anything* that involves extraterrestrial intelligence—has been ordained as a very highly-classified, secret subject, the consequence of which we would see as a total blackout on public research and discussion by official science bodies such as NASA.

And this is essentially what we have today.

Well, I suppose in a world where profit-driven success dominates virtually everything that moves upon the Earth we should not be surprised at the thought of 'outer space and aliens' too being the subject of power, control, competition, and commercial opportunity. If true, these corporations and covert groups would have to have immense influence amongst established and legitimate government agencies, for such control would surely demand a level of cooperation at the highest level. A blind eye turned this way and that. The promise of remuneration perhaps.

Is this why elected governments have done virtually nothing about revealing this information to the people of the world? Too many public individuals are in too deep?

One could explore further into why many prominent politicians belong to various secret orders and societies and consequently, may have mysterious agendas that do not befit their office and oath, and whole books have indeed been written about how deep the rabbit hole goes in this regard and how and why world affairs take the course they do. But while corrupt politicians and leaders may be prevalent in our world today, there are also a great many concerned people out there who have become genuinely frustrated with the complete lack of engagement by governments on the subject of extraterrestrials, and who want things to change.

Witness the words of **John Podesta**, **former President Bill Clinton's chief of staff and counselor to President Obama**:[98]

> It's time to open the books on questions that have remained in the dark, questions of government investigations of UFOs. We ought to do it because the American

> people quite frankly can handle the truth. And we ought to do it because it's the law.

What frustrations prompted this high-level politician to make such a public declaration? Has he perhaps tried to uncover information on this subject, but his efforts have been met with a brick wall?

Former deputy prime minister and Canadian minister of national defense, Paul Hellyer gives us a clue. He believes that a United States 'shadow government' has used extraterrestrial technology at secret 'black op' locations. He said:[99]

> I believe that they have developed energy sources, and publicly I'm saying that if they do not exist in commercial form, that extraterrestrials would certainly give us that information if we would ask them for it and stop shooting at them.

The implication of this comment is staggering. Hellyer is saying that a rogue, well-financed group in America is in control of creating new energy sources stolen from recovered alien vehicles. But not only that—that they are also actively engaged in trying to shoot down even more E.T. craft to steal more technology.

The other surely profound implication here is the idea that friendly extraterrestrial civilizations are willing to give us new energy sources to help us out of our energy crisis. And we can see now where this is going.

Do our energy companies really want aliens to present us with a clean, cheap energy source that anyone can use and not pay a fortune for? And if the oil, gas, coal, and nuclear companies can steal that cheap energy source, patent it and then charge us all for using it when the fossil fuels run out…

Yes, I believe the rabbit hole goes very deep indeed.

It is perhaps not surprising then that throughout recent history many prominent citizens have voiced their opinion regarding the existence and reality of extraterrestrial visitation to our planet, and yet, even their significant and remarkable statements have fallen largely on deaf ears.

Former CIA director, Vice Admiral Roscoe Hillenkoetter:

> Unknown objects are operating under intelligent control . . . It is imperative that we learn where UFOs come from and what their purpose is.[100]

> It is time for the truth to be brought out in open congressional hearings. Behind the scenes, high-ranking Air Force officers are soberly concerned about the UFOs. But through official secrecy and ridicule, many citizens are led to believe the unknown flying objects are nonsense.[101]

Victor Marchetti, former executive assistant to the deputy director of the CIA:[102]

> My theory is that we have, indeed, been contacted—perhaps even visited—by extraterrestrial beings, and the U.S. government, in collusion with the other national powers of the Earth, is determined to keep this information from the general public.

Air Chief Marshal Lord Dowding, **commander-in-chief of the Royal Air Force Fighter Command during the Battle of Britain:**[103]

> I am convinced that these objects do exist and that they are not manufactured by any nation on Earth. I can, therefore, see no alternative to accepting the theory that they come from some extraterrestrial source.

Mikhail Gorbachev, former president of the USSR:[104]

> The phenomenon of UFOs does exist, and it must be treated seriously.

Dr. Margaret Mead Ph.D., famous anthropologist:[105]

> We can only imagine what purpose lies behind the activities of these quiet, harmlessly cruising objects that time and again approach the Earth. The most likely explanation, it seems to me, is that they are simply watching what we are up to.

C. G. Jung, eminent analytical psychologist:[106]

> A purely psychological explanation is ruled out ... the discs show signs of intelligent guidance, by quasi-human pilots . . . the authorities in possession of important information should not hesitate to enlighten the public as soon and as completely as possible.

Professor Hermann Oberth, German rocket expert who worked with Wernher von Braun, the Army Ballistic Missile Agency and later NASA:[107]

> It is my thesis that flying saucers are real and that they are space ships from another solar system. I think that they possibly are manned by intelligent observers who are members of a race that may have been investigating our Earth for centuries. I think that they possibly have been sent out to conduct systematic, long-range investigations, first of men, animals, vegetation, and more recently of atomic centers, armaments, and centers of armament production.

As a rocket scientist Professor Oberth was deeply interested in the propulsion systems of UFOs:[108]

> They are flying by means of artificial fields of gravity... They produce high-tension electric charges in order to push the air out of their paths, so it does not start glowing, and strong magnetic fields to influence the ionized air at higher altitudes. First, this would explain their lu-

minosity; secondly, it would explain the noiselessness of UFO flight.

Taking all of the above statements together we can be in no doubt at all that we have a pretty serious, very real, and unequivocal situation at hand: extraterrestrial spaceships are 100% real and are visiting the Earth and studying humanity.

And what do we get from science on this deep and profound subject?

Absolutely nothing.

Most of today's scientists won't come anywhere near UFO research. It is regarded by them as a toxic subject that will harm their careers, their scientific reputation, and their funding. And yet, why do we have scientists at all if they won't look at the evidence for the most important contact that the human race is ever likely to make? Contact with an extraterrestrial civilization.

When trained astronauts, airline and military pilots, air traffic controllers—people who are highly trained technicians we trust with our very lives, on a day-to-day basis—tell us something is visiting our world that is not from this Earth, why is this testimony not sufficient persuasion for our scientists to become deeply engaged, do their duty and study this matter for the sake of humanity?

Something is very wrong here. Is it the same blinkered idleness that won't allow them to look at the evidence for a civilization on Mars, because 'they don't believe it's scientifically possible'? Or is it the system itself that is corrupted to ensure this subject is never investigated?

Dr. Jacques Vallee, astrophysicist, computer scientist, researcher and author on UFOs, has a scathing word to say about such scientific failings:[109]

> Skeptics, who flatly deny the existence of any unexplained phenomenon in the name of 'rationalism,' are

among the primary contributors to the rejection of science by the public. People are not stupid and they know very well when they have seen something out of the ordinary. When a so-called expert tells them the object must have been the moon or a mirage, he is really teaching the public that science is impotent or unwilling to pursue the study of the unknown.

Dr. J. Allen Hynek, a university astronomer and scientific consultant on UFOs to the U.S. Air Force from 1948 until 1969 also had this to say about the limits of scientific perspective, in a letter to *Science Magazine*:[110]

> I have begun to feel that there is a tendency in 20th Century science to forget that there will be a 21st Century science and indeed a 30th Century science, from which vantage points our knowledge of the universe may appear quite different than it does to us. We suffer, perhaps, from temporal provincialism, a form of arrogance that has always irritated posterity.

The only explanation that makes any sense is if the orders come from the very top. At the highest political level, or from a level that may exist beyond that, there must be some agreement that there is to be no public acknowledgment of the existence of UFOs and extraterrestrial civilizations visiting the Earth.

From an illuminating article entitled "How the CIA Views the UFO Phenomenon"[111] Vincent Marchetti, former CIA and author of the book *The CIA and the Cult of Intelligence,* the only book ever censored by the U.S. Government prior to publication up to 1979, explained why he, from an intelligence perspective, believed the UFO cover-up existed:

> The purpose of the international conspiracy is to maintain a workable stability among the nations of the world and for them, in turn, **to retain institutional control over their respective populations**.

> Thus, **for these governments to admit there are beings from outer space attempting to contact us**, beings with mentalities and technological capabilities obviously far superior to ours, **could, once fully perceived by the average person, erode the foundations of the Earth's traditional power structure**.

Marchetti continues his sobering and pragmatic assessment:

> Political and legal systems, religions, economic and social institutions could all soon become meaningless in the mind of the public. The national oligarchical establishments, even civilization as we know it, could collapse into anarchy. **Such extreme conclusions are not necessarily valid, but they probably accurately reflect the fears of the "ruling class"** of the major nations, whose leaders (particularly those in the intelligence business) have always advocated excessive governmental secrecy as being necessary to preserve "national security."
>
> **The real reason for such secrecy is, of course, to keep the public uninformed, misinformed, and, therefore, malleable**. [Emphasis added]

Marchetti's view is one born of many years experience in a major international intelligence agency and is surely likely to be very close to the truth, particularly as he is said to have left the CIA somewhat disillusioned with the service. He presents a damning indictment of those who we assume serve our best interests in wanting to preserve society and civilization, but who are simply doing what they think is necessary to preserve their power base and control.

However, for such a worldwide 'gagging order' to be effective, there must be the means to limit press coverage and of course scientific curiosity—because there will be many intelligent reporters and scientists who *will* be aware of this mass of evidence. Is it possible for worldwide news about UFOs to be curtailed and science to be restrained from investigating and examining hugely significant matters?

Regarding global news coverage, it is well known that the world's media is controlled by a tiny handful of international corporations, so it would not be too great a difficulty to imagine an accord existing between shady government offices, media barons, and the intelligence services. The only question being, why would the corporate leaders agree to something like that, and what would they want in exchange for their silence and co-operation?

Of course, there are always some reports of UFOs in the media, TV documentaries on the subject, and of course many major movies on the topic too. But you may have noticed a very common trend between all of this public material:

- **The news reports** *always* portray UFO evidence with a humorous, 'little green men' not-to-be-taken-seriously slant, accompanied by grinning newsreaders and spooky X-Files-themed music playing in the background

- **The documentaries** *always* leave the question open as to whether the evidence and the witnesses can be believed or not

- **The movies**...well, when was the last science fiction film made that did *not* feature aliens intent on destroying the Earth and humanity?

If memory serves, I think the last movie made about peaceful aliens was Steven Spielberg's *E.T. the Extra-Terrestrial* over 30 years ago. A disturbing trend in itself, which would suggest that someone somewhere wants humanity to believe that aliens are a dangerous threat and will go to some length to make sure that only certain types of alien movies get bankrolled for Hollywood.

Regarding the scientists, we have seen that they largely self-regulate themselves through peer review and reputation, so the fear of embarrassment and being discredited is a strong motivator for avoiding UFO research. Their scientific pursuits will also be curtailed by whoever funds their grants, and certainly, NASA does not currently fund any UFO research, which leaves us with the question:

Who exactly is directing the secrecy program and liaising with NASA to ensure the matter of 'discoveries of extraterrestrial intelligence' is kept under wraps and the public kept firmly in the dark?

At the level of 'photo doctoring,' the American NSA (National Security Agency) would appear to be involved, at least back in the 1960s, but as I have no knowledge of how these agencies interact with each other and compartmentalize security I'm not going to attempt to work out who ends up being responsible for what, and to be quite honest, I really don't care and we are unlikely to discover those answers anyway.

All that matters is this: at what level has the American government *lost control of the oversight of contact with extraterrestrial civilizations*—or indeed, does it still retain control of this domain and is, therefore, lying to the world about its knowledge?

No doubt, at some stage 'plausible deniability' has come into play so that the president and elected government officials can deny any knowledge of extraterrestrial contact and don't have to say or confirm anything to the public. But as we now see, given the huge weight of evidence and testimony for this contact having been made many, many times over decades, the U.S. government's position has been rendered quite ridiculous, farcical, and completely unbelievable.

And if, as is becoming apparent, rogue, shadowy groups have achieved control over extraterrestrial matters that belong in the hands of the elected government and the people, then they know what must be done.

But will they do anything?

In 2011 a petition from the American public initiated by the Paradigm Research Group asked the Obama administration to "Officially acknowledge an extraterrestrial presence engaging the human race and to release all of their files"—an opportunity perhaps for the president to come clean about the truth and to end decades of lies and disinformation to the people.

But in response to this petition, the official statement issued by the White House through their Office of Science & Technology Policy,[112] revealed just how unreachable, impotent and truly out of touch with the people they are—a people who clearly know a lot more than they do on this matter.

It reads:

> The U.S. government has no evidence that any life exists outside of our planet, or that an extraterrestrial presence has contacted or engaged any member of the human race. In addition, **there is no credible information to suggest that any evidence is being hidden from the public's eye**.

And then further states:

> The odds of us making contact with any of them—especially intelligent ones—are extremely small, given the distances involved. The fact is **we have no credible evidence of extraterrestrial presence here on Earth**. [Emphasis added]

Considering all we have presented in this book, all the impeccable, validated testimony, and analysis given by highly credible individuals and witnesses of extraterrestrial events, it is incredulous that such a witless, inane and quite frankly stupid declaration could ever be uttered by an elected government to its people, and at that, be expected to be believed.

What was it that Victor Marchetti, the Executive Assistant to the Deputy Director of the CIA said? "The real reason for such secrecy is, of course, to keep the public uninformed, misinformed, and, therefore, malleable."

The people of this world need to awaken to the true reality of life on Earth and in this universe. Step up the courageous world leaders we now need who are willing and able to take this on.

Those who truly proffer to serve the people.

Chapter 6: Conclusion

We began this book by asking why NASA had sent five successive landing spacecraft to Mars with not one astrobiology experiment on board that could detect life and also, why it had avoided landing in locations where liquid water was known to exist, which would have afforded the opportunity to search for life if they had wanted to do so.

We then wondered why NASA had ignored evidence from their image libraries, which showed possible artificial objects and structures on the surface of Mars that pointed to the possibility of a civilization having once existed there, while too assessing evidence elsewhere such as on our own Moon, that showed signs of an intelligent extraterrestrial presence that had also been ignored.

We then explored how this problem of obfuscation of valid questions concerning scientific research on Mars extended to the general position of denial that government and the scientific community has with the entire question regarding intelligent extraterrestrials visiting our solar system and humanity, a position utterly in conflict with the impeccable testimony and expert opinion declaring their existence to be an undeniable fact.

I believe that all of this evidence shows that NASA is curtailed by a secrecy policy that forbids acknowledgment and discussion of the matter of intelligent signs of E.T. life—alien life—whether this has been found to exist on other planets in our solar system or whether it is evident from visitations to the Earth from other civilizations in our galaxy, or beyond it.

They have managed to achieve and maintain this clandestine position through hiding behind a posture of scientific legitimacy, in that the only subjects concerning extraterrestrial matters that are ever discussed and presented to the eyes and minds of the public and the media are via the strictly limited fields of mainstream planetary science, i.e.

geology, climate studies, and primitive life/microbiology. Any talk of aliens and intelligent life in the galaxy is confined to discussing the severely limited scope of SETI and their search for radio signals in space, or to the more respectable Kepler Mission that is scanning for Earth-like worlds in the cosmos.

There can be no doubt that all evidence of extraterrestrial intelligence is officially being kept secret at the highest level, and knowledge of its existence completely denied to the public. This can be seen as plainly obvious by the simple fact that NASA's astronauts, who would be considered to be the epitome of experience and commonsense in the matter of space and what is out there, are at odds with the official position adopted by NASA and the United States government, which claims there is 'no evidence' for intelligent extraterrestrial life, past or present, in our solar system.

It is an untenable position. The evidence will continue to grow, the frustrations of the public will eventually build to a roar, and commonsense will eventually prevail and the truth will come out. The process by which it will unfold is perhaps what we are seeing now—a gradual trickle of acknowledgment by respected persons such as astronauts; the release of UFO files by government departments; the Vatican making proclamations on the likely existence of extraterrestrial life—slow and gradual acclimatization for the masses.

But how far will it go? The real test will come if information comes to light concerning the really 'extreme' subjects such as, for example, the recovery of crashed and back-engineered alien craft we have mentioned; events that have been substantially reported upon and which are supported by highly credible and reliable testimony. If detailed information such as this is released, ever more serious questions would then be prompted, such as:

- Where are those craft being kept now, and when can the public see them?

- What was done with the technology discovered in the craft and who has benefitted from it?
- What was learned from the inhabitants and where are they now?
- Where did the craft's inhabitants come from?
- Why are they visiting the Earth?

The questions will keep coming, and they won't stop until all of the truth is out. And this may be why so little has been revealed so far. It could be that the pressure is mounting on the authorities to release 'something', to distract and appease the public, but that they don't want to give away too much. They don't want to upset the applecart and perhaps all of the dubious investments that have been made during the preceding decades of secrecy.

It seems to me that someone, somewhere, in some powerful capacity still wants to retain a strong measure of control over the release of this information.

But will we stand for that? Are we going to allow them to continue to deny humanity and every one of us the truth about life on Earth and about those sentient beings that are making the effort to visit us from far distant stars, or from dimensions in space that we don't yet understand? What do our leaders know, and why haven't they told us about these visitors?

Are we going to sit back and allow them to keep from us the incredible knowledge of possible ancient civilizations that may once have existed in our solar system, on the Planet Mars, and the Moon? Are we indeed the offspring of ancient colonization from an extraterrestrial race, as some of the evidence in this book suggests?

These are questions that demand the whole of humanity be engaged with so that we can learn about our true history as a human species. Such knowledge does not belong hidden, in a secret vault. It is the heritage of humanity. We cannot allow a select and privileged few to secretly study and learn the true nature and extent of life in the universe so that they can then decide how best to position themselves to make use of this information, while we, the rest of humanity, wander around in ignorance and at their behest.

Final Thoughts

The images I have shown in this book offer some fascinating and intriguing evidence that the remains of an ancient civilization exist on the planet Mars, the Moon, and possibly throughout our entire solar system. The possibility that these worlds are still inhabited by intelligent beings and are being used as bases, is supported too by the massive activity going on in the present day regarding visitations to the Earth by what are almost certainly, extraterrestrial or extradimensional craft.

Because this compelling evidence is completely ignored by NASA and the U.S. government, I and many other researchers have concluded that a deliberate policy must exist to withhold this information from the world and to keep it secret. This book is, therefore, a challenge to the authorities to come clean to the people and to open up all the files on this subject—in effect, to initiate full disclosure of what is known about intelligent extraterrestrial life and our contact with it.

It is not lost on me that this is a massive subject to engage with. As a mere layman with a deep interest in this matter, I have done my best to present to you the most outstanding evidence I have seen so far, as an overview and introduction to this subject. It is for others far more qualified to take this extremely important study further to open humanity's eyes to the true nature of what our exploration of Mars and our solar system might be revealing to us.

I'll be straight up with you. I don't think there is one single piece of image evidence here that is by itself, indisputable proof of buildings, habitats, constructs, or small artifacts that belong to a Mars or Moon civilization. However, I do strongly believe that taking all of this evidence together *as a whole*, we do have a very serious case for these signs of extraterrestrial intelligence existing for real, whether they originate in the ancient past or, possibly, still exist today.

I believe the evidence in this book shows that NASA should actively and openly be looking for the remains of possible Mars and Moon civilizations and that they should be employing archaeologists to study the orbital images for signs of buried buildings and habitats and large-scale technology. Although now, of course, we have a pretty good idea as to why they are not doing so, at least publicly.

But of one thing you can be sure: this information is known and is deliberately being concealed from us. More significantly, this knowledge is being denied to the entire human race, and we are entitled to know why.

In Summary

Well, that's it. If the evidence in this book has moved you to question what you believe you know about the world, then it has done its work.

Whether you are a scientist, an investigative journalist, or a sovereign being of this planet simply wanting to know the truth, you are now in a position to make things happen. I think you should be up in arms about it and demanding answers from your political leaders to such questions as:

- Who is gaining from these expensive space missions to the planets and beyond, if the more important discoveries being made are withheld from the public?

- What is being done with this secret knowledge that you and I are not allowed to know about?

- Are publicly funded space missions being hijacked by a self-serving clique of corporate industrialists who plan to commercially exploit discoveries of extraterrestrial technology?

- Is the reality of E.T. causing a panic in the echelons of power? Are they afraid they will lose their control over humanity if people across the world learn about these beings and the true history of our solar system?

In closing, I would like to say that I am fully prepared for every single piece of evidence presented here to be challenged, and my conclusions proved wrong. All I am asking is that this evidence gets exposed, studied, and assessed properly. Because if I am right and there is a secrecy agenda being pursued regarding our scientific investigations on Mars and elsewhere in our solar system concerning the suppression of evidence for intelligent extraterrestrial life, then we have a very serious moral, ethical and philosophical problem deeply rooted in our world, and it needs to be dealt with urgently and at the highest level.

The real question, however, is this: "When are we, humanity, a race of conscious intelligent beings, going to grow up?"

Can we really consider ourselves civilized when we routinely suppress important knowledge discovered during our journeys into space . . . knowledge and truth that is of profound significance to the whole of the human race?

If we want this situation to change, however, and if our politicians and leaders won't take the necessary steps, then it will be left to us, the people, to take up the challenge.

I hope and trust that what has been revealed in these pages proves to be the catalyst for yet another brick in the wall to work its way loose, hit the ground, and shatter. And that a wiser and freer humanity will duly have cause to awaken, on this precious Earth.

Wishing profound peace to you.

M. J. Craig

Epilogue

One day a cataclysm may destroy the Earth and most traces of human civilization will be wiped from the surface of the planet. For many thousands of years, and much longer, the echoes of human life may lay unheard as Earth's oceans dry up and disappear and vast wastelands of desert take their place as the sole legacy of our once beautiful world.

On another planet in our solar system, a young civilization emerges into their space-age and starts to send robotic vehicles to explore the red, dying world that we once called Earth. Images returned by their orbiting spacecraft reveal that the planet once had river valleys and oceans, a very long time ago, and could, therefore, have possibly harbored life.

They land a spacecraft at a location that may have been the site of a port, as studies show that a sea would have once lapped this shoreline and high-resolution images reveal signs of possible artificial structures there too, although geologists dismiss them as remnant massifs and mesas, of which there are many thousands to be seen.

A robotic rover lands and explores the area. Its camera examines the ground carefully. It happens upon a broken plate, a rusted gear mechanism, and a square metal box. Their scientists immediately declare to their world that this represents serious, unambiguous, and incontrovertible evidence that intelligent life once existed on the Red Planet. They immediately begin to re-appraise their history as a civilization and contemplate their place in the universe…

This is how science should be progressing on Mars. But we are not doing it. If we are going to roam about on other planets with expensive mobile laboratories, dig up soil to analyze for microscopic life, but pretend we didn't see that child's shoe lying on the ground a few feet away . . . who is fooling who? If we find just one single fork, spoon, or tea plate on the surface of Mars or the Moon, we have to ask:

"How did it get there?"

And if we don't, and just ignore it, then why in heaven's name are we spending billions going to that planet in the first place?

Someone is trying to fool you and I, that much is certain.

M J C

References, Notes & Info

Research

• **Unusual Martian Surface Features**, V. DiPietro/G. Molenaar, Mars Research, (1982)

• **The Monuments of Mars**, R. C. Hoagland, 4th Ed, Frog, (1996)

• **The Martian Enigmas: A Closer Look**, M. J. Carlotto, North Atlantic, (1991)

• **Hoagland's Mars Vol. 2: The UN Briefing**, (video) R. C. Hoagland, (1992),
http://www.enterprisemission.com/videos.html

• **Hoagland's Mars Vol. 3: The Moon-Mars Connection** (video) R. C. Hoagland, (1994)
http://www.enterprisemission.com/videos.html

• **The McDaniel Report**, S. V. McDaniel, North Atlantic, (1993)

• **The Case for the Face**, Ed. S. V. McDaniel, M. Paxson, Adventures Unlimited, (1998)

• **A Passion for Mars**, A. Chaikin, Abrams, (2008)

• **Dark Mission: The Secret History of NASA**, R. C. Hoagland, M. Bara, 2nd Ed, Feral House, (2009)

• **The Hidden Truth: Water and Life on Mars**, J. P. Skipper, Planetary Publishing, (2010),
http://www.marsanomalyresearch.com/general-directories/book-1/ad-doc.htm

• **Extraterrestrial Contact: The Evidence and Implications**, S. M. Greer, M.D., Crossing Point, (1999)

• **Disclosure: Military and Government Witnesses Reveal the Greatest Secrets in Modern History** (kindle eBook), S. M. Greer M.D., Crossing Point, (2001)

• **Proposed Studies on the Implications of Peaceful Space Activities for Human Affairs**, D. N. Michael, Brookings Institute (1960), ttps://www.brookings.edu/blog/brookings-now/2014/05/12/communications-technology-and-extraterrestrial-life-the-advice-brookings-gave-nasa-about-the-space-program-in-1960/

• **Disclosure Project: 2001 National Press Club Event** (video) https://player.vimeo.com/video/88355599

• **The Citizen Hearing on Disclosure of an Extraterrestrial Presence Engaging the Human Race** (videos)
http://www.citizenhearing.org

Website Links

• **Richard C. Hoagland: The Enterprise Mission**
http://www.enterprisemission.com

• **Joseph P Skipper: Mars Anomaly Research**
http://www.marsanomalyresearch.com

• **Keith Laney**
http://thehiddenmission.com

• **Mark J. Carlotto (Digital Enhancement of Martian Anomalies)**
http://carlotto.us/Martianenigmas/index.shtml

• **The Society for Planetary SETI Research**
http://spsr.utsi.edu

• **Efrain Palermo: Martian Dark Stains**
http://palermoproject.com/Martian_Stains.html

•**Dr. Steven Greer/The Disclosure Project:** Disclosing the Facts about UFOs, Extraterrestrial Intelligence, and Classified Advanced Energy and Propulsion Systems
http://www.disclosureproject.org

• **The Citizen Hearing on Disclosure**
http://www.citizenshearing.org

Notes

Chapter 1: Obfuscation

[1] **Desert Varnish on Mars** (article), http://www.astrobio.net/mars/desert-varnish-on-mars/
[2] **AOSS professors aid in search for life on Mars** (article), https://www.michigandaily.com/article/u-professors-are-vital-members-new-mars-mission
[3] **The Steppenwolf: A Proposal for a Habitable Planet in Interstellar Space** (paper), http://arxiv.org/abs/1102.1108
[4] **The Science of Curiosity: Seeking Signs of Past Mars Habitability** (video), http://www.jpl.nasa.gov/video/details.php?id=1095
[5] **Mars Astrobiology Field Lab Rover: In Depth** (web pg), https://solarsystem.nasa.gov/missions/mafl/indepth
[6] **Circadian Rhythms and Evidence for Life on Mars** (paper),http://www.researchgate.net/publication/228422463_Circadian_Rhythms_and_Evidence_for_Life_on_Mars
[7] **The Monuments of Mars** (book), Richard C. Hoagland, 4th Ed, Frog, (1996), p5
[8] **The Geomorphology and Geometry of the D&M Pyramid** (paper), Erol O. Torun, http://kbmorgan.com/MarsMission/marsbull/bull201.htm
[9] **Unusual Martian Surface Features** (book), V. DiPietro and G. Molenaar, Mars Research (1982)
[10] **Hoagland's Mars Vol. 2: The Terrestrial Connection: The UN Briefing** (video), Richard C. Hoagland, http://www.enterprisemission.com/videos.html
[11] **The Mounds of Cydonia: A Case Study for Planetary SETI** (paper), Horace W. Crater, http://spsr.utsi.edu/
[12] **The Martian Enigmas: A Closer Look** (book), Mark J. Carlotto, North Atlantic, (1991)
[13] **The McDaniel Report** (book), Stanley V. McDaniel, North Atlantic, (1993) p3
[14] **The McDaniel Report . . .** p10

[15] **A Passion for Mars** (book), Andrew Chaikin, Abrams (2008) p204
[16] **The McDaniel Report** (book), Stanley V. McDaniel, North Atlantic, (1993) pv
[17] **The McDaniel Report . . .** p69
[18] **Proposed Studies on the Implications of Peaceful Space Activities for Human Affairs** (book), https://www.brookings.edu/blog/brookings-now/2014/05/12/communications-technology-and-extraterrestrial-life-the-advice-brookings-gave-nasa-about-the-space-program-in-1960/
[19] **UFOs Are as Real As the Airplanes Flying Overhead, says Former Defense Minister Paul Hellyer** (video), https://www.youtube.com/watch?v=fjYLI_y_3SM
[20] **The Disclosure Project: A Research Project Working to Fully Disclose the Facts about UFOs, Extraterrestrial Intelligence, and Classified Advanced Energy and Propulsion Systems** (website), http://www.disclosureproject.org
[21] **Understanding UFO Secrecy** (article), Steven M. Greer M.D., p2 http://www.disclosureproject.org/docs/pdf /UnderstandingUFOSecrecy.pdf
[22] **President Eisenhower's Farewell Address** (speech), https://en.wikipedia.org/wiki/Eisenhower's_farewell_address

Chapter 2: Water on Mars
[23] **NASA Counts Down the Hours to its Latest Mission: is there Life on Mars?** (article), https://www.theguardian.com/science/2012/jul/14/nasa-mission-to-mars
[24] **Mars Science Laboratory Launch** (press kit), p10 http://www.jpl.nasa.gov/news/press_kits/MSLLanding.pdf
[25] **A Passion for Mars** (book), Andrew Chaikin. p208
[26] **NASA's Next Mars Rover to Land at Gale Crater** (article), http://www.nasa.gov/mission_pages/msl/news/msl20110722.html
[27] **Water on Mars Confirmed by Hoagland** (article),

http://www.enterprisemission.com/kelp.htm
[28] **Evidence for Recent Groundwater Seepage and Surface Runoff on Mars** (paper), M. Malin/K. Edgett http://www.sciencemag.org/site/feature/data/hottopics/se2600 02330p.pdf
[29] **Mars Map Projection Plotting Approximate Locations of MOC Stain Images** (web pg), http://palermoproject.com/Mars_Anomalies/MarsStainMap.html
[30] **Martian Water Stains or Dust Slides?** (paper), E. Palermo/ J. England/H. Moore
http://palermoproject.com/SeepsPaper.pdf
[31] **Flowing Water May Exist on Mars** (article), http://physicsworld.com/cws/article/news/2011/aug/04/flowi ng-water-may-exist-on-mars
[32] **The Hidden Truth: Water & Life on Mars** (book), J. P. Skipper, Planetary Publishing (2010) p155
[33] **NASA Spacecraft Data Suggest Water Flowing On Mars** (article), http://mars.jpl.nasa.gov/mro/news /whatsnew/index.cfm?FuseAction=ShowNews&NewsID=1144
[34] **Physical and Thermodynamical Evidence for Liquid Water on Mars** (paper),
http://www.lpi.usra.edu/meetings/lpsc2009/pdf/1440.pdf
[35] **University Researchers Discovers Liquid Saltwater on Mars** (article), http://www.michigandaily.com/content /2009-04-02/u-professor-discovers-liquid-salt-water-mars
[36] **Don Juan Pond, Antarctica: Near-surface CaCl2-brine Feeding Earth's Most Saline Lake and Implications for Mars** (article), http://www.nature.com/articles/srep01166
[37] **NASA Mission Overview** (web page), http://mars.jpl.nasa. gov/programmissions/overview
[38] **NASA Weighs Use of Rover to Image Potential Mars Water Sites** (article), https://www.nasa.gov/feature/nasa-weighs-use-of-rover-to-image-potential-mars-water-sites

[39] **Searching for Life in Martian Water Will Be Very, Very Tricky** (article), *h*ttp://www.scientificamerican.com/article/searching-for-life-in-martian-water-will-be-very-very-tricky/
[40] **Scientific Objectives of the ExoMars Rover** (web pg), http://exploration.esa.int/mars/45082-rover-scientific-objectives/
[41] **Should We Search For Life On Mars Before Sending Astronauts?: Unfortunately, NASA Doesn't Have a Good Game Plan** (article), http://www.popsci.com/we-should-probably-check-for-life-on-mars-before-we-send-astronauts-there
[42] **Martians Might Be Real. That Makes Mars Exploration Way More Complicated** (article), http://www.wired.com/2016/08/shouldnt-go-mars-might-decimate-martians/#slide-1
[43] See note 37
[44] **Drill Bits on Rover Could Contaminate Mars** (article), http://www.post-gazette.com/science/2012/09/16/Drill-bits-on-rover-could-contaminate-Mars/stories/201209160221
[45] **Send Astronauts to Mars to Find Evidence of Life, NASA's Top Scientist Says** (article), http://www.space.com/32909-mars-life-search-nasa-astronauts.html
[46] **NASA's Humans-to-Mars Plans Win Publicity But Lack Details** (article), http://spacenews.com/nasas-humans-to-mars-plans-win-publicity-but-lack-details/

Chapter 3: Signs of Life on Mars

[47] **Mars Gone Wild** (article), https://www.wired.com/2004/08/mars-5/
[48] **The Hidden Truth: Water & Life on Mars** (book), J. P. Skipper p.157
[49] **Bone up on Mars Rock Shapes** (web pg), http://mars.jpl.nasa.gov/msl/multimedia/images/?ImageID=6538
[50] **Mars' Atmosphere Was Likely More Oxygen-Rich Long Ago** (article),

http://www.space.com/33296-mars-atmosphere-oxygen-curiosity-rover.html
[51] **The Curious Case of the NASA Crinoid Cover-Up** (article), http://www.enterprisemission.com/_articles/03-08-2004/crinoid_cover-up.htm
[52] **Former NASA Scientist Claims Conspiracy About Mars Photo** (video), https://www.youtube.com/watch?v=6E7aqCaekDA
[53] **Potential Signs of Ancient Life in Mars Rover Photos** (article), http://www.astrobio.net/news-exclusive/potential-signs-ancient-life-mars-rover-photos/
[54] **Ancient Sedimentary Structures in the <3.7 Ga Gillespie Lake Member, Mars, That Resemble . . .** (paper), http://online.liebertpub.com/doi/abs/10.1089/ast.2014.1218?journalCode=ast
[55] **Mars Microbe Traces Spotted by Rover? Probably Not, Curiosity Team Says** (article), http://www.space.com/28218-mars-rover-curiosity-signs-life.html
[56] **Alien Crab Fake** (web pg), http://thehiddenmission.com/wholelotoffaking/goingon/aliencrabfoundonmars.html
[57] **NASA Curiosity Raw Image Database** (web pg), http://mars.nasa.gov/msl/multimedia/raw/
[58] **Teeny Little Bigfoot on Mars** (article), http://www.planetary.org/blogs/emily-lakdawalla/2008/1305.html

Chapter 4: Signs of Civilization on Mars

[59] **Opportunity Making Its Way to Final Position on Cape Desire: Sol-by-Sol Summary** (web pg), http://www.jpl.nasa.gov/missions/mer/images-print.cfm?id=2112
[60] **The Hidden Truth: Water & Life on Mars** (book), J. P. Skipper, p255
[61] **The Hidden Truth: Water & Life on Mars . . .** p257
[62] **Life After People** (video), http://topdocumentaryfilms.com/life-after-people/

[63] **Mars Reconnaissance Orbiter and 'The Lost Cities of Barsoom'** (article), http://www.enterprisemission.com/ Lost-CitiesofBarsoom.htm
[64] See note 10.
[65] **ESA Disclaimer:** Image artifacts and artificially-looking features are often the results of atmospheric distortion which are caused by either dust or clouds and haze. If an image appears blurry in certain areas or even noisy but the color representation is comparable to that in cleaner areas, there is good chance that there is some local dust. If color representation of certain regions is remarkably different from other areas, i.e. faint blue or white coverage, there is a good chance that haze or clouds have been imaged. This is, however, not always the case and it depends largely on the area context that has been imaged. Areas at generally low elevations such as Valles Marineris, Hellas and Argyre Planitiae, or the Northern Lowlands, for example, behave in a different way than areas at high elevations such as the volcanic edifices.
HRSC color filters are not identical to what the human eye would identify as RGB, so color calculations and representations depend on the area size per image that has been disturbed by atmospheric effects. Pan-sharpening techniques using a combination of the color channels and the panchromatic channel can also pronounce such areas, especially if they appear only locally. Effects might show up in e.g. the blue channel while the nadir and other color channels are clean. This way, color scenes might look a bit different from the nadir greyscale scene.
[66] **The Light Finally Dawns at Cydonia** (article), http://www.enterprisemission.com/paper_1/paper_1.html
[67] **The Craters are Electric** (article), Michael Goodspeed https://www.thunderbolts.info/webnews/120707electriccraters.htm

Chapter 5: The Alien Connection

[68] **UFOs and Defence: What Should We Prepare For?** (report), http://www.ufoevidence.org/topics/cometa.htm
[69] **Disclosure Project: 2001 National Press Club Event** *(video)*, https://player.vimeo.com/video/88355599
[70] **Disclosure: Military and Government Witnesses Reveal the Greatest Secrets in Modern History** (kindle eBook), Steven M. Greer M.D., Crossing Point, (2001), loc 853
[71] **Artificial Satellites of Mars/Riddle of the Martian Satellites: I. S. Shklovskiy** (interview transcript),

http://earthsbanner.com/shklovskii/Shklovskii.html
[72] **For the World is Hollow, and I Have Touched the Sky** (article), http://www.enterprisemission.com/Phobos.html
[73] **For the World is Hollow and I Have Touched the Sky, Pt2** http://www.enterprisemission.com/Phobos2.html
[74] **Palermo's Phobos Anomalies** (web pg), http://palermoproject.com/Mars_Anomalies/PhobosAnomalies1.html
[75] **Apollo, 11 25th Anniversary—The White House** (video), https://www.youtube.com/watch?v=Znyx2gTh3HU
[76] **Hoagland's Mars Vol. 3: The Moon-Mars Connection** (video), http://www.enterprisemission.com/videos.html
[77] **Dark Mission** (book), R. C. Hoagland/M. Bara, Feral house, 2007 p206-210
[78] **Dark Mission . . .** p215-17
[79] **Disclosure Project: 2001 National Press Club Event** (video), https://player.vimeo.com/video/88355599
[80] **The NASA Conspiracy: Donna Hare** (video), https://www.youtube.com/watch?v=tEBLmWhx1K0
[81] **Apollo 17 Lunar Surface Journal: Traverse to Geology Station 3** (audio & transcript), Time ref: 143:50:54 https://www.hq.nasa.gov/alsj/a17/a17.trvsta3.html
[82] **Apollo 17 Lunar Surface Journal: Geology Station 3 at Ballet Crater** (audio & transcript), Time ref: 145:23:3 https://www.hq.nasa.gov/alsj/a17/a17.trvsta4.html
[83] **Dark Mission** (book), R. C. Hoagland/M. Bara, pg 560-61, Figs.27-28
[84] **Apollo 17 Lunar Surface Journal: Orange Soil** (audio & transcript), https://www.hq.nasa.gov/alsj/a17/a17.sta4.html
[85] See note 80
[86] **The Citizen Hearing on Disclosure of an Extraterrestrial Presence Engaging the Human Race** (videos), http://www.citizenhearing.org
[87] **PRG Quotes: Dr. Edgar Mitchell** (web pg), http://www.paradigmresearchgroup.org/QuotesPage.htm#M

[88] **PRG Quotes: Cmdr. Eugene Cernan** (web pg), http://www.paradigmresearchgroup.org/QuotesPage.htm#C
[89] See note 87
[90] **Thrive Movement: UFOs—Brian O'Leary** (video), https://www.youtube.com/watch?v=yO0T05kQkbs
[91] **Gordon Cooper's Letter to Grenada U.N. Ambassador** http://www.ufoevidence.org/news/article161.htm
[92] **PRG Quotes: Col. L. Gordon Cooper** (web pg), http://www.paradigmresearchgroup.org/QuotesPage.htm#C
[93] **Story Musgrave Talks About Extraterrestrials** (video), https://www.youtube.com/watch?v=rL-EOPvuXqk
[94] **Extraterrestrial Contact : The Evidence and Implications** (book), Steven M. Greer M.D, Crossing Point, p314-15
[95] **Disclosure: Military and Government Witnesses Reveal the Greatest Secrets in Modern History** (kindle eBook), Steven M. Greer M.D., loc 931
[96] **Disclosure: Military and Government . . .** loc 6551
[97] **Disclosure: Military and Government . . .** loc 1035
[98] **Obama Aide John Podesta on UFOs and alien Encounters** (video), https://www.youtube.com/watch?v=YKwFP7ZcDwY
[99] **Paul Hellyer, Light at the End of the Tunnel** (interview transcript), http://projectavalon.net/lang/en/paul_hellyer_light_at_the_end_of_the_tunnel_en.html
[100] **UFO Briefing Document: The Best Available Evidence** (Kindle eBook), Don Berliner, Dell (2000), loc 553
[101] **Air Force Order on 'Saucers' Cited** (article), http://query.nytimes.com/mem/archive/pdf?res=9A00E3D81F39EF32A2575BC2A9649C946191D6CF
[102] **How the CIA Views the UFO Phenomenon** (article), http://www.theufochronicles.com/2015/02/how-cia-views-ufo-phenomenon.html
[103] **PRG Quotes: Air Chief Marshal Lord Dowding** (web pg), http://www.paradigmresearchgroup.org/QuotesPage .htm#D
[104] **PRG Quotes: President Gorbachev** (web pg), http://www.paradigmresearchgroup.org/QuotesPage.htm#G

[105] **Disclosure: Military and Government Witnesses Reveal the Greatest Secrets in Modern History** (kindle eBook), Steven M. Greer M.D., loc 845
[106] **Disclosure: Military and Government . . .** loc 889
[107] **Disclosure: Military and Government . . .** loc 885
[108] **UFO Briefing Document: The Best Available Evidence** (Kindle eBook), Don Berliner, loc 2778
[109] **UFO Briefing Document: The Best Available . . .** loc 2694
[110] **UFO Briefing Document: The Best Available . . .** loc 303
[111] See note 99
[112] **Searching for E.T., But No Evidence Yet : White House Response to Petition** (Web pg), https://petitions.whitehouse.gov/petition/formally-acknowledge-extraterrestrial-presence-engaging-human-race-disclosure

Indexes

Using the Image Index

- **How to find the images**—the easiest way to see the original space agency images is **NOT** to try and copy the web links listed here (which will be frustrating), but instead to go to our website at www.secretmars.com. Go to the *Secret Mars Book* menu and click on the *Image evidence* page. These will match the "SM numbers" as shown here in the book i.e. "**SM1. The Face on Mars: original image**". The reason I've included the links in this book is for posterity, as one day the website and I will disappear.

- **Finding the anomalous objects**—most of the anomalies will easily be found on the image, some may take a minute or so. But it is well worth the time to view the original pictures on your computer screen where much more detail will be seen than it is possible to reproduce here in this printed book. At the space agency websites, many images will also be in color (the Kindle edition of this book is in color).

- **Gallery**—where this word appears, I am citing the official web page where several thumbnail images are stored of images related to the main one. Useful if you want to dig deeper.

- **Weblinks not working?**—finally, I should mention that NASA and its affiliates often shift their databases around and cause havoc (I'm sure it's not intentional). This may sometimes result in a web link no longer working. However, the most common reason why a link won't work is when you painstakingly try to copy a link from this book and invariably miss a digit, or just type it wrong . . . *(believe me, I know!)*

Image Index 1: Secret Mars Image Evidence

- **SM1. The Face on Mars: original image**

M. J. Carlotto enhancement:
http://carlotto.us/martianenigmas/Articles/vikFace/vikface.shtml
NASA original:
http://photojournal.jpl.nasa.gov/catalog/PIA01141

- **SM2. The Face on Mars: second image**

http://carlotto.us/martianenigmas/Articles/vikFace/vikface.shtml

- **SM3. The D&M Pyramid: Viking Mission**

http://carlotto.us/martianenigmas/Articles/vikDM/vikDM.shtml

- **SM4. The "City" with the triangular-shaped "Fort"**

M. J. Carlotto enhancement:
http://carlotto.us/martianenigmas/Articles/April_2000/April2000.shtml
NASA original:
http://nssdc.gsfc.nasa.gov/image/planetary/mars/f035a72_processed.jpg

- **SM5. Water flow from a crater wall**

https://ida.wr.usgs.gov/fullres/divided/sp2338/sp233806a.jpg

- **SM6, 7 & 8. Examples of "Dark flows"**

http://ida.wr.usgs.gov/fullres/divided/m08076/m0807686b.jpg
http://ida.wr.usgs.gov/fullres/divided/m09020/m0902083d.jpg
http://ida.wr.usgs.gov/fullres/divided/m02047/m0204738e.jpg

- **SM9. Dark flows from Arabia Terra Region**

http://hirise.lpl.arizona.edu/ESP_022405_1910

- **SM10. Spring and summer flows, Newton Crater**

http://www.nasa.gov/mission_pages/MRO/multimedia/pia14479.html

- **SM11. Liquid water flow in Gale Crater**

http://mars.nasa.gov/msl/multimedia/raw/?rawid=0707ML0030050000304601E01_DXXX&s=707
'Curiosity rover driving past the water':
https://www.universetoday.com/wp-content/uploads/2014/08/Curiosity-Sol-711_3a_Ken-Kremer1.jpg

- **SM12. A lake of liquid water on Mars?**

https://ida.wr.usgs.gov/fullres/divided/m09013/m0901354a.jpg

- **SM13. Lakes of ice or liquid water?**

https://ida.wr.usgs.gov/fullres/divided/m09020/m0902042d.jpg

- **SM14. More lakes?**

https://ida.wr.usgs.gov/fullres/divided/e09003/e0900304f.jpg

- **SM15. Group of lakes or polar pits?**

https://ida.wr.usgs.gov/fullres/divided/e09000/e0900020a.jpg

- **SM16. "Dark Lake"**

https://ida.wr.usgs.gov/fullres/divided/e08009/e0800954a.jpg

• **SM17. "Lakes and islets"**
https://ida.wr.usgs.gov/fullres/divided/e08010/e0801033b.jpg
• **SM19. Possible frozen water is analyzed. . .**
Upper image:
http://mars.nasa.gov/msl/multimedia/raw/?rawid=1349ML0064810010600346E01_DXXX&s=1349
Lower image:
http://mars.jpl.nasa.gov/msl-raw-images/msss/01350/mcam/1350MR0064880000701088E02_DXXX.jpg
• **SM20. Trees on Mars? NASA description. . .**
https://ida.wr.usgs.gov/fullres/divided/m08046/m0804688a.jpg
• **SM21. "Martian Spiders"**
http://HiRISE.lpl.arizona.edu/ESP_020914_0930
• **SM22. "Starburst Spiders"**
http://hirise-pds.lpl.arizona.edu/PDS/EXTRAS/RDR/ESP/ORB_011800_011899/ESP_011842_0980/ESP_011842_0980_RGB.NOMAP.browse.jpg
• **SM23. Starburst Spiders in the springtime. . .**
http://hirise-pds.lpl.arizona.edu/PDS/EXTRAS/RDR/ESP/ORB_023100_023199/ESP_023117_0980/ESP_023117_0980_RGB.NOMAP.browse.jpg
• **SM24. "Alpine scenery"**
https://ida.wr.usgs.gov/fullres/divided/e07017/e0701717b.jpg
• **SM25. "Martian bushes"**
https://ida.wr.usgs.gov/fullres/divided/s06006/s0600607a.jpg
• **SM26. NASA: "Defrosting dark spot"**
https://ida.wr.usgs.gov/fullres/divided/e09003/e0900320b.jpg
• **SM27. Martian vegetation or geology?**
https://ida.wr.usgs.gov/fullres/divided/s07026/s0702623b.jpg
• **SM28. Martian "Petri dish"**
https://ida.wr.usgs.gov/fullres/divided/m08000/m0800063c.jpg
• **SM29. Martian skull?**
http://marsrover.nasa.gov/gallery/all/2/p/513/2P171912249EFFAAL4P2425L7M1.HTML
• **SM30. Animal skeleton?**
http://mars.nasa.gov/mer/gallery/press/spirit/20061025a/McMurdo_L257F-A814R1.jpg
• **SM31. Fossil field?**
http://marsrovers.jpl.nasa.gov/gallery/all/2/p/016/2P127793693EFF0327P2371R1M1.HTML
• **SM32. Fossilized thigh bone?**

http://mars.nasa.gov/msl/multimedia/images/?ImageID=6538

• **SM33. Skull and Teeth?**

http://mars.jpl.nasa.gov/msl-raw-images/msss/00107/mcam
/0107MR0682050000E1_DXXX.jpg

• **SM34. Animal Skeleton?**

http://mars.jpl.nasa.gov/msl-raw-
images/msss/00109/mcam/0109MR0684021000E1_DXXX.jpg

• **SM35. Ammonite fossil?**

http://mars.jpl.nasa.gov/msl-raw-
images/msss/00518/mhli/0518MH0261000000E1_DXXX.jpg
Ammonite fossil from Earth:
http://www.sedgwickmuseum.org/index.php?page=coastal-plains

• **SM36. Possible crinoid fossil destroyed by NASA?**

Image before destruction:
http://mars.nasa.gov/mer/gallery/all/1/m/034/1M131201699EFF05
00P2933M2M1.HTML
Image after destruction:
http://mars.nasa.gov/mer/gallery/all/1/m/034/1M131212854EFF05
00P2959M2M1.HTML
Gallery:
http://mars.nasa.gov/mer/gallery/all/opportunity_m034.html

• **SM37. Debris field?**

http://marsrovers.jpl.nasa.gov/gallery/all/2/p/527/2P173156766EFF
ACA0P2440R1M1.HTML
Color panorama:
http://photojournal.jpl.nasa.gov/catalog/pia04182

• **SM38. Junkyard artifact?**

http://marsrovers.jpl.nasa.gov/gallery/all/2/p/015/2P127699373EFF
0313P2549L7M1.JPG

• **SM39. Damaged container oozing liquid...?**

Image 12D091.BB2, CE label:
http://pds-imaging.jpl.nasa.gov/vikingl/vl_images.html
Research by T. Beech/J. P. Skipper:
http://www.marsanomalyresearch.com/evidence-reports/2009/163
/moving-evidence.htm

•**SM40. Metal pipe or life form?**

http://mars.nasa.gov/mer/gallery/all/2/p/229/2P146694689EFF860
0P2402L7M1.JPG

• **SM41. NASA debris or what?**

http://marsrovers.jpl.nasa.gov/gallery/all/2/n/036/2N129561872EF
F 0361P1617R0M1.JPG

• **SM42. The "Spider-crab" of Mars**

http://mars.jpl.nasa.gov/msl-raw-images/msss/00710/mcam/0710MR0030150070402501E01_DXXX.jpg

• **SM43. The "Ghost lady" of Mars**

http://mars.jpl.nasa.gov/msl-raw-images/msss/01001/mcam/1001ML0044610000305331D01_DXXX.jpg

Another image:

http://mars.jpl.nasa.gov/msl-raw-images/proj/msl/redops/ods/ surface/sol/01001/opgs/edr/ncam/NRB_486375691EDR_ F048157 0NCAM00322M_.JPG

Gallery:

http://mars.nasa.gov/msl/multimedia/raw/?s=#/?slide=1001

• **SM44. The "Statuette lady" of Mars**

http://photojournal.jpl.nasa.gov/jpeg/PIA10216.jpg

• **SM45. The "Gear wheel"**

http://marsrovers.nasa.gov/gallery/all/2/p/288/2P151930534EFF8987P2418R1M1.HTML

• **SM46. Metal box or casing?**

http://marsrovers.jpl.nasa.gov/gallery/all/2/p/1419/2P252334617EFFAX00P2260L2M1.JPG

• **SM47. Manufactured object?**

http://marsrovers.jpl.nasa.gov/gallery/all/2/p/1402/2P250825588EFFAW9DP2432R1M1.HTML

• **SM48. Small coin?**

http://photojournal.jpl.nasa.gov/catalog/PIA01907

• **SM49. The "Wheel hub"**

http://mars.jpl.nasa.gov/msl-raw-images/msss/00064/mcam/0064MR0285005000E1_DXXX.jpg

• **SM50. The "Tiny turbine"**

http://mars.jpl.nasa.gov/msl-raw-images/msss/01000/mcam/1000MR0044630400503600E02_DXXX.jpg

• **SM51. Small pyramid**

http://mars.jpl.nasa.gov/msl-raw-images/msss/00978/mcam/0978MR0043250040502821E01_DXXX.jpg

• **SM52. The "C"**

http://www.jpl.nasa.gov/spaceimages/images/largesize/PIA19066_hires.jpg

• **SM53. The "Door bolt"**

http://mars.jpl.nasa.gov/msl-raw-images/msss/00440/mcam/0440MR1795005000E2_DXXX.jpg

Another image:

http://mars.jpl.nasa.gov/msl-raw-images/msss/00440/mcam/0440ML0017950060201701E01_DXXX.jpg

• **SM54. Wheels and axle?**
http://mars.jpl.nasa.gov/msl-raw-images/msss/00729/mcam/0729
ML0031250020305133E01_DXXX.jpg
• **SM55. The "Nozzle"**
http://mars.jpl.nasa.gov/msl-raw-images/msss/00821/mcam/0821
MR0036170080500530E01_DXXX.jpg
• **SM56. Masonry carving?**
http://mars.jpl.nasa.gov/msl-raw-images/msss/00991/mcam/0991
ML0043820010404528E01_DXXX.jpg
• **SM57. Beam clamp strut?**
http://mars.jpl.nasa.gov/msl-raw-images/msss/00528/mcam/0528
ML0020870010203210E01_DXXX.jpg
Panorama of "Dingo Gap":
http://photojournal.jpl.nasa.gov/catalog/PIA17931
• **SM58. Mechanical fitting?**
http://mars.jpl.nasa.gov/msl-raw-images/msss/00109/mcam/0109
MR0684024000E1_DXXX.jpg
• **SM59. Stanchion base plate?**
http://mars.jpl.nasa.gov/msl-raw-images/msss/00595/mcam/0595
MR0025090390400725E01_DXXX.jpg
• **SM60. Crank handle device?**
http://mars.jpl.nasa.gov/msl-raw-images/msss/01051/mcam/1051
MR0046240040104587E01_DXXX.jpg
Another image:
http://mars.jpl.nasa.gov/msl-raw-images/msss/01049/mcam/1049
ML0046190030306063E01_DXXX.jpg
• **SM62. Martian cairn?**
http://marsrovers.nasa.gov/gallery/all/1/p/2467/1P347194239EFFB
0Q0 P2366L6M1.HTML
• **SM63. Building foundation?**
Composite image sources:
Image 1:
http://mars.jpl.nasa.gov/msl-raw-images/msss/00529/mcam/0529
MR0020960070303356E01_DXXX.jpg
Image 2:
http://mars.jpl.nasa.gov/msl-raw-images/msss/00529/mcam/0529
MR0020960080303357E01_DXXX.jpg
• **SM64. Natural or artificial mounds?**
https://ida.wr.usgs.gov/fullres/divided/m00016/m0001661c.jpg
• **SM65. Trail left by the rover, or something else?**
http://marsrovers.nasa.gov/gallery/all/1/p/1070/1P223169173EFF7
8VAP2629L6M1.JPG

• **SM66. Large scale pipe constructions?**
https://ida.wr.usgs.gov/fullres/divided/m11000/m1100099d.jpg
• **SM67. Martian tunneling mechanism?**
https://ida.wr.usgs.gov/display/MGSC_1096/m12004/m1200441.imq.jpg
• **SM68. Ancient Martian ruins?**
https://ida.wr.usgs.gov/display/MGSC_1184/e10004/e1000462.imq.jpg
Research by R C Hoagland:
http://www.enterprisemission.com/LostCitiesofBarsoom.htm
• **SM69. Partially buried remains of a Martian village?**
http://photojournal.jpl.nasa.gov/jpeg/PIA08014.jpg
(This is a large image and the location is difficult to find, so here's some help...)

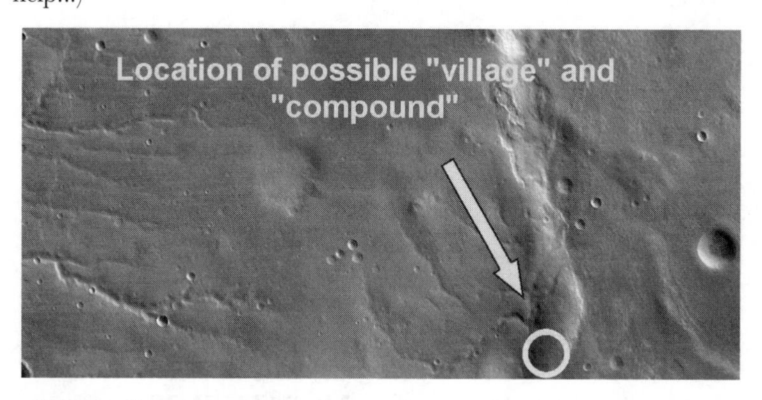

• **SM70. Geometric shapes**
https://ida.wr.usgs.gov/fullres/divided/ab1084/ab108405e.jpg
• **SM70.1 Geometric shapes — HiRISE/MGS comparison**
HiRISE image (top)
http://hirise-pds.lpl.arizona.edu/PDS/EXTRAS/RDR/ESP/ORB_011300_011399/ESP_011372_1730/ESP_011372_1730_RED.NOMAP.browse.jpg
• **SM71. Cydonia: possible artificial structures**
http://spaceinimages.esa.int/Images/2006/09/Cydonia_region_color_image
• **SM72. The Face on Mars: Viking/HiRISE comparison**
Viking:
http://photojournal.jpl.nasa.gov/catalog/PIA01141
HiRISE:
http://hirise-pds.lpl.arizona.edu/PDS/EXTRAS/RDR/PSP/ ORB_003200_003299/PSP_003234_2210/PSP_003234_2210_RED.abrowse.jpg
• **SM73. The Face on Mars: 2001**

http://www.msss.com/mars_images/moc/extended_may2001/face/face_E03-00824_proc.gif

- **SM74—75. The Face on Mars: 2006 — Mars Express**

Overhead image:
http://spaceinimages.esa.int/Images/2006/09/Cydonia_region_color_image
Perspective image:
http://spaceinimages.esa.int/Images/2006/09/Face_on_Mars_in_Cydonia_region_perspective2

- **SM76. The Cydonia Region: 2006 — Mars Express**

http://spaceinimages.esa.int/Images/2006/09/Cydonia_region_color_image

- **SM77. The Fort: comparison 1976-2006**

(L) Viking Orbiter:
http://nssdc.gsfc.nasa.gov/image/planetary/mars/f035a72_processed.jpg
(Centre) MGS Orbiter:
https://ida.wr.usgs.gov/display/MGSC_1044/m04019/m0401903.imq.jpg
M. J. Carlotto analysis:
http://carlotto.us/martianenigmas/Articles/April_2000/April2000.shtml
(R) Mars Express:
http://spaceinimages.esa.int/Images/2006/09/Cydonia_region_color_image

- **SM78. The "Five" — Mars Express 2006**

http://spaceinimages.esa.int/Images/2006/09/Cydonia_region_color_image

- **SM79. The Five — HiRISE/MRO 2008**

http://viewer.mars.asu.edu/planetview/inst/ctx/P15_006992_2194_XN_39N009W#P=P15_006992_2194_XN_39N009W&T=2

- **SM80. The D&M Pyramid — MGS/THEMIS**

http://photojournal.jpl.nasa.gov/jpegMod/PIA04745_modest.jpg

- **SM82. The Hexagon — Mars Express 2006**

http://spaceinimages.esa.int/Images/2006/09/Cydonia_region_color_image

- **SM83. The Hexagon — THEMIS/Mars Odyssey 2002**

http://photojournal.jpl.nasa.gov/jpegMod/PIA04057_modest.jpg

- **SM84. The "H" Structure — Mars Express 2006**

http://spaceinimages.esa.int/Images/2006/09/Cydonia_region_color_image

- **SM86. Phobos: structure revealed. . .**

http://solarsystem.nasa.gov/multimedia/display.cfm?Category=Planets&IM_ID=6103

- **SM87. Phobos: rectilinear grid. . .**

http://www.esa.int/var/esa/storage/images/esa_multimedia/images/2010/03/mars_moon_phobos/9916979-2-eng-GB/Mars_moon_Phobos.tif

- **SM88. The Phobos monolith — MOC/MGS Orbiter**

https://ida.wr.usgs.gov/fullres/divided/sp2551/sp255103h.jpg

• **SM89. The Iapetus wall — Cassini Orbiter 2004**
http://photojournal.jpl.nasa.gov/catalog/PIA06166
• **SM90. The geometric horizon of Iapetus. . .**
http://photojournal.jpl.nasa.gov/jpegMod/PIA06146_modest.jpg
Research & analysis by R. C. Hoagland:
http://www.enterprisemission.com/moon2.htm
• **SM91. The "Castle" discovered by R. C. Hoagland**
Images & analysis of the "Castle":
http://www.tarrdaniel.com/documents/Ufology/lunar_anomaly.html
Official Apollo image showing "No Castle":
http://www.lpi.usra.edu/resources/apollo/frame/?AS10-32-4822
• **SM92. Lunar junkyard at Shorty Crater? — Apollo 17**
http://history.nasa.gov/alsj/a17/AS17-137-21000HR.jpg
• **SM93. Artificial object at Henry Crater?**
https://www.hq.nasa.gov/alsj/a17/AS17-140-21409HR.jpg
Another image, different perspective:
https://www.hq.nasa.gov/alsj/a17/AS17-140-21485HR.jpg
Research & analysis by R. C. Hoagland:
http://www.enterprisemission.com/datashead.htm
• **SM94. More artifacts near Henry Crater?**
https://www.hq.nasa.gov/alsj/a17/AS17-141-21582HR.jpg

Image Index 2: Additional Images

- **1. Probable Martian meteorite: ALH84001**
Image credits: NASA/JPL
- **2. Possible fossilized Martian nanobacteria. . .**
Image credit: NASA/JPL
- **3. "Dust Devils" forming elaborate tracks. . .**
Image credit: NASA/JPL/University of Arizona
http://hirise-pds.lpl.arizona.edu/PDS/EXTRAS/RDR/ESP/ORB_014400_014499/ESP_014426_2070/ESP_014426_2070_RGB.NOMAP.browse.jpg
- **4. Streaks of dust sliding down dune slopes. . .**
Image credit: NASA/JPL/University of Arizona
http://hirise.lpl.arizona.edu/images/2009/details/cut/PSP_007962_2635_cut.jpg
- **5. The Mars Science Laboratory (MSL) — Artist's. . .**
Image credit: NASA/JPL-Caltech
- **6. Viking Lander 2 on Mars: Utopian Plain, 1976**
Image credit: NASA/JPL
- **7. Mars Observer: reported lost on approach. . .**
Image credit: NASA/JPL
- **8. Desolate Mars. . .**
Image credit: NASA/JPL/Cornell
http://photojournal.jpl.nasa.gov/jpeg/PIA10216.jpg
- **9.** Channels carved by water: Nirgal Vallis. . .
Image credit: NASA/JPL
- **10. Teardrop-shaped 'islands' in Ares Vallis**
Image credit: NASA/JPL
- **11. Archaea, halobacteria . . .**
Image credit: NASA
- **12. Mars Phoenix Lander: water droplets. . .**
Image credit: NASA/JPL-Caltech/U. of Arizona/Max Planck Inst.
- **13. Lake Urmia: Iran, Earth**
Image credit: NASA
- **14. Water lake: South America, Earth**
Image credit: ©2013 DigitalGlobe, Google Earth
- **15. Water lake: South America, Earth**
Image credit: ©2013 DigitalGlobe, Google Earth
- **16. Don Juan Pond. . .**
Image credit: Samantha Joye
- **17. A Viking Lander—about to be baked. . .**
Image credit: NASA/JPL

• 18. Spring begins in the Martian south polar. . .
Image credit: Arizona State University/Ron Miller
• 19. Porous volcanic rock
Image credit: NASA/JPL/Cornell
http://mars.nasa.gov/mer/gallery/press/spirit/20061025a/McMurdo_L257F-A814R1.jpg
• 20. Fossilized thigh bone of a bison
Image credit: public domain
• (SM35). Ammonite fossil? — Curiosity Rover
(Lower right) © 2016 Sedgwick Museum of Earth Sciences, U. of Cambridge (reproduced with permission)
http://www.sedgwickmuseum.org/index.php?page=coastal-plains
• 21. Rock at Gillespie Lake: ancient home for. . .
Image credit: NASA/JPL-Caltech/MSSS
http://mars.nasa.gov/msl/multimedia/raw/?rawid=0126MR0007820030200790E01_DXXX&s=126
• 22. The Spirit Rover: Gusev Crater — Artist's. . .
Image credit: NASA/JPL
• 23. MSL Curiosity at Gale Crater — (Selfie)
Image credit: NASA/JPL-Caltech
• 24. Was all hope for life evolving on Mars. . .
Image credit: iStock.com
• 25. The Pyramids in Egypt
Image credit: iStock.com
• 26. Extensive geometrical patterns reveal. . .
Image credit: NASA/Landsat
• 27. Machu Picchu — Peru, South America
Image credit: iStock.com
• 28. Earth ruins, near Sarvistan, Iran — 1936
Image credit: Oriental Institute, University of Chicago
Slide No.30:
http://oi.uchicago.edu/gallery/pa_iran_paai_asf/index.php/10C7_72dpi.png?action=big&size=original
• (SM69.4). Town ruins in Ashur, Iraq
Image credit (L): ©2013 DigitalGlobe, Google Earth
• 29. Phobos: the largest Martian moon. . .
Image credit: NASA/JPL/University of Arizona
http://solarsystem.nasa.gov/multimedia/display.cfm?Category=Planets&IM_ID=6103
• 30. Iapetus: natural or artificial moon? Cassini. . .
Image credit: NASA/JPL/Space Science Institute

http://photojournal.jpl.nasa.gov/catalog/PIA11690
- **31. Astronaut Buzz Aldrin: Stepping onto. . .**

Image credit: NASA/JPL
- **32. Lunar Rover Vehicle at Shorty Crater. . .**

Source image credits: NASA/JPL; composite picture: M. J. Craig
Source images used for composite:
https://www.hq.nasa.gov/alsj/a17/AS17-137-21009HR.jpg
https://www.hq.nasa.gov/alsj/a17/AS17-137-20994HR.jpg
https://www.hq.nasa.gov/alsj/a17/AS17-137-20993HR.jpg

Book Index

About the Author

M. J. Craig is a writer, researcher, and composer involved in projects to support a more enlightened civilization on our planet. His first book *Secret Mars: the Alien Connection* was published in 2013, followed by a revised and expanded edition in 2017 and a translation into German (*Geheimer Mars*). His latest book *Memories of Mars* is due in 2020.

For correspondence to the author, please email:

contact@secretmars.com

For information and updates, visit the website at:

www.secretmars.com

ANCIENT ALIENS ON THE MOON
By Mike Bara

What did NASA find in their explorations of the solar system that they may have kept from the general public? How ancient really are these ruins on the Moon? Using official NASA and Russian photos of the Moon, Bara looks at vast cityscapes and domes in the Sinus Medii region as well as glass domes in the Crisium region. Bara also takes a detailed look at the mission of Apollo 17 and the case that this was a salvage mission, primarily concerned with investigating an opening into a massive hexagonal ruin near the landing site. Chapters include: The History of Lunar Anomalies; The Early 20th Century; Sinus Medii; To the Moon Alice!; Mare Crisium; Yes, Virginia, We Really Went to the Moon; Apollo 17; more. Tons of photos of the Moon examined for possible structures and other anomalies.
248 Pages. 6x9 Paperback. Illustrated.. $19.95. Code: AAOM

ANCIENT ALIENS ON MARS
By Mike Bara

Bara brings us this lavishly illustrated volume on alien structures on Mars. Was there once a vast, technologically advanced civilization on Mars, and did it leave evidence of its existence behind for humans to find eons later? Did these advanced extraterrestrial visitors vanish in a solar system wide cataclysm of their own making, only to make their way to Earth and start anew? Was Mars once as lush and green as the Earth, and teeming with life? Chapters include: War of the Worlds; The Mars Tidal Model; The Death of Mars; Cydonia and the Face on Mars; The Monuments of Mars; The Search for Life on Mars; The True Colors of Mars and The Pathfinder Sphinx; more. Color section.
252 Pages. 6x9 Paperback. Illustrated. $19.95. Code: AMAR

ANCIENT ALIENS ON MARS II
By Mike Bara

Using data acquired from sophisticated new scientific instruments like the Mars Odyssey THEMIS infrared imager, Bara shows that the region of Cydonia overlays a vast underground city full of enormous structures and devices that may still be operating. He peels back the layers of mystery to show images of tunnel systems, temples and ruins, and exposes the sophisticated NASA conspiracy designed to hide them. Bara also tackles the enigma of Mars' hollowed out moon Phobos, and exposes evidence that it is artificial. Long-held myths about Mars, including claims that it is protected by a sophisticated UFO defense system, are examined. Data from the Mars rovers Spirit, Opportunity and Curiosity are examined; everything from fossilized plants to mechanical debris is exposed in images taken directly from NASA's own archives.
294 Pages. 6x9 Paperback. Illustrated. $19.95. Code: AAM2

ANCIENT TECHNOLOGY IN PERU & BOLIVIA
By David Hatcher Childress

Childress speculates on the existence of a sunken city in Lake Titicaca and reveals new evidence that the Sumerians may have arrived in South America 4,000 years ago. He demonstrates that the use of "keystone cuts" with metal clamps poured into them to secure megalithic construction was an advanced technology used all over the world, from the Andes to Egypt, Greece and Southeast Asia. He maintains that only power tools could have made the intricate articulation and drill holes found in extremely hard granite and basalt blocks in Bolivia and Peru, and that the megalith builders had to have had advanced methods for moving and stacking gigantic blocks of stone, some weighing over 100 tons.
340 Pages. 6x9 Paperback. Illustrated.. $19.95 Code: ATP

ANTARCTICA AND THE SECRET SPACE PROGRAM
By David Hatcher Childress

David Childress, popular author and star of the History Channel's show *Ancient Aliens*, brings us the incredible tale of Nazi submarines and secret weapons in Antarctica and elsewhere. He then examines Operation High-Jump with Admiral Richard Byrd in 1947 and the battle that he apparently had in Antarctica with flying saucers. Through "Operation Paperclip," the Nazis infiltrated aerospace companies, banking, media, and the US government, including NASA and the CIA after WWII. Does the US Navy have a secret space program that includes huge ships and hundreds of astronauts?

392 Pages. 6x9 Paperback. Illustrated. $22.00 Code: ASSP

NORTH CAUCASUS DOLMENS
By Boris Loza, Ph.D.

Join Boris Loza as he travels to his ancestral homeland to uncover and explore dolmens firsthand. Chapters include: Ancient Mystic Megaliths; Who Built the Dolmens?; Why the Dolmens were Built; Asian Connection; Indian Connection; Greek Connection; Olmec and Maya Connection; Sun Worshippers; Dolmens and Archeoastronomy; Location of Dolmen Quarries; Hidden Power of Dolmens; and much more! Tons of Illustrations! A fascinating book of little-seen megaliths. Color section.

252 Pages. 5x9 Paperback. Illustrated. $24.00. Code NCD

THE ENCYCLOPEDIA OF MOON MYSTERIES
Secrets, Anomalies, Extraterrestrials and More
By Constance Victoria Briggs

Our moon is an enigma. The ancients viewed it as a light to guide them in the darkness, and a god to be worshipped. Did you know that: Aristotle and Plato wrote about a time when there was no Moon? Several of the NASA astronauts reported seeing UFOs while traveling to the Moon?; the Moon might be hollow?; Apollo 10 astronauts heard strange "space music" when traveling on the far side of the Moon?; strange and unexplained lights have been seen on the Moon for centuries?; there are said to be ruins of structures on the Moon?; there is an ancient tale that suggests that the first human was created on the Moon?; Tons more. Tons of illustrations with A to Z sections for easy reference and reading.

152 Pages. 7x10 Paperback. Illustrated. $19.95. Code: EOMM

OBELISKS: TOWERS OF POWER
The Mysterious Purpose of Obelisks
By David Hatcher Childress

Some obelisks weigh over 500 tons and are massive blocks of polished granite that would be extremely difficult to quarry and erect even with modern equipment. Why did ancient civilizations in Egypt, Ethiopia and elsewhere undertake the massive enterprise it would have been to erect a single obelisk, much less dozens of them? Were they energy towers that could receive or transmit energy? With discussions on Tesla's wireless power, and the use of obelisks as gigantic acupuncture needles for earth, Chapters include: Megaliths Around the World and their Purpose; The Crystal Towers of Egypt; The Obelisks of Ethiopia; Obelisks in Europe and Asia; Mysterious Obelisks in the Americas; The Terrible Crystal Towers of Atlantis; Tesla's Wireless Power Distribution System; Obelisks on the Moon; more. 8-page color section.

336 Pages. 6x9 Paperback. Illustrated. $22.00 Code: OBK

ANCIENT ALIENS AND JFK
The Race to the Moon & the Kennedy Assassination
By Mike Bara

Relying on never-before-seen documents culled from the recent Kennedy assassination papers document dump, Bara shows the secret connections between key assassination figures like Oswald, LBJ, and highly placed figures inside NASA who had reasons to want Kennedy dead. Bara also looks into the bizarre billion-dollar Treasury bonds that Japanese businessmen attempted to deposit in a Swiss bank that had photos of Kennedy and the Moon on them. Is the wealth of the Moon being used as collateral by the USA? The book will dig deeply into Kennedy's silent war with shadowy Deep State figures who were desperate to shut down his Disclosure agenda. Also: the Apollo mission; "Apollo 20," and more. Includes 8-page color section.

248 Pages. 6x9 Paperback. Illustrated. $19.95. Code: AAJK

HIDDEN AGENDA
NASA and the Secret Space Program
By Mike Bara

Bara delves into secret bases on the Moon, and exploring the many other rumors surrounding the military's secret projects in space. On June 8, 1959, a group at the ABMA produced for the US Department of the Army a report entitled Project Horizon, a "Study for the Establishment of a Lunar Military Outpost." The permanent outpost was predicted to cost $6 billion and was to become operational in December 1966 with twelve soldiers stationed at the Moon base. Does hacker Gary Mackinnon's discovery of defense department documents identifying "non-terrestrial officers" serving in space? Includes an 8-page color section.

346 Pages. 6x9 Paperback. Illustrated. $19.95. Code: HDAG

THE ANTI-GRAVITY FILES
A Compilation of Patents and Reports
Edited by David Hatcher Childress

In the tradition of *The Anti-Gravity Handbook* and *the Time-Travel Handbook* comes this compilation of material on anti-gravity, free energy, flying saucers and Tesla technology. With plenty of technical drawings and explanations, this book reveals suppressed technology that will change the world in ways we can only dream of. Chapters include: A Brief History of Anti-Gravity Patents; The Motionless Electromagnet Generator Patent; Mercury Anti-Gravity Gyros; The Tesla Pyramid Engine; Anti-Gravity Propulsion Dynamics; The Machines in Flight; More Anti-Gravity Patents; Death Rays Anyone?; The Unified Field Theory of Gravity; and tons more. Heavily illustrated. 4-page color section.

216 pages. 8x10 Paperback. Illustrated. $22.00. Code: AGF

PROJECT MK-ULTRA AND MIND CONTROL TECHNOLOGY
By Axel Balthazar

This book is a compilation of the government's documentation on MK-Ultra, the CIA's mind control experimentation on unwitting human subjects, as well as over 150 patents pertaining to artificial telepathy (voice-to-skull technology), behavior modification through radio frequencies, directed energy weapons, electronic monitoring, implantable nanotechnology, and more.

384 pages. 7x10 Paperback. Illustrated. $19.95. Code: PMK

ORDER FORM

10% Discount When You Order 3 or More Items

One Adventure Place
P.O. Box 74
Kempton, Illinois 60946
United States of America
Tel.: 815-253-6390 • Fax: 815-253-6300
Email: auphq@frontiernet.net
http://www.adventuresunlimitedpress.com

ORDERING INSTRUCTIONS

- ✓ Remit by USD$ Check, Money Order or Credit Card
- ✓ Visa, Master Card, Discover & AmEx Accepted
- ✓ Paypal Payments Can Be Made To: info@wexclub.com
- ✓ Prices May Change Without Notice
- ✓ 10% Discount for 3 or More Items

SHIPPING CHARGES

United States

- ✓ Postal Book Rate { $4.50 First Item; 50¢ Each Additional Item
- ✓ POSTAL BOOK RATE Cannot Be Tracked! Not responsible for non-delivery.
- ✓ Priority Mail { $7.00 First Item; $2.00 Each Additional Item
- ✓ UPS { $9.00 First Item (Minimum 5 Books); $1.50 Each Additional Item

NOTE: UPS Delivery Available to Mainland USA Only

Canada

- ✓ Postal Air Mail { $19.00 First Item; $3.00 Each Additional Item
- ✓ Personal Checks or Bank Drafts MUST BE US$ and Drawn on a US Bank
- ✓ Canadian Postal Money Orders OK
- ✓ Payment MUST BE US$

All Other Countries

- ✓ Sorry, No Surface Delivery!
- ✓ Postal Air Mail { $19.00 First Item; $7.00 Each Additional Item
- ✓ Checks and Money Orders MUST BE US$ and Drawn on a US Bank or branch.
- ✓ Paypal Payments Can Be Made in US$ To: info@wexclub.com

SPECIAL NOTES

- ✓ RETAILERS: Standard Discounts Available
- ✓ BACKORDERS: We Backorder all Out-of-Stock Items Unless Otherwise Requested
- ✓ PRO FORMA INVOICES: Available on Request
- ✓ DVD Return Policy: Replace defective DVDs only

ORDER ONLINE AT: www.adventuresunlimitedpress.com

10% Discount When You Order 3 or More Items!

Please check: ☑

☐ This is my first order ☐ I have ordered before

Name

Address

City

State/Province | Postal Code

Country

Phone: Day | Evening

Fax | Email

Item Code	Item Description	Qty	Total

Subtotal ▶

Please check: ☑

Less Discount-10% for 3 or more items ▶

☐ Postal-Surface — Balance ▶

☐ Postal-Air Mail (Priority in USA) — Illinois Residents 6.25% Sales Tax ▶

Previous Credit ▶

☐ UPS (Mainland USA only) — Shipping ▶

Total (check/MO in USD$ only) ▶

☐ Visa/MasterCard/Discover/American Express

Card Number:

Expiration Date: Security Code:

✓ SEND A CATALOG TO A FRIEND: